T0138317

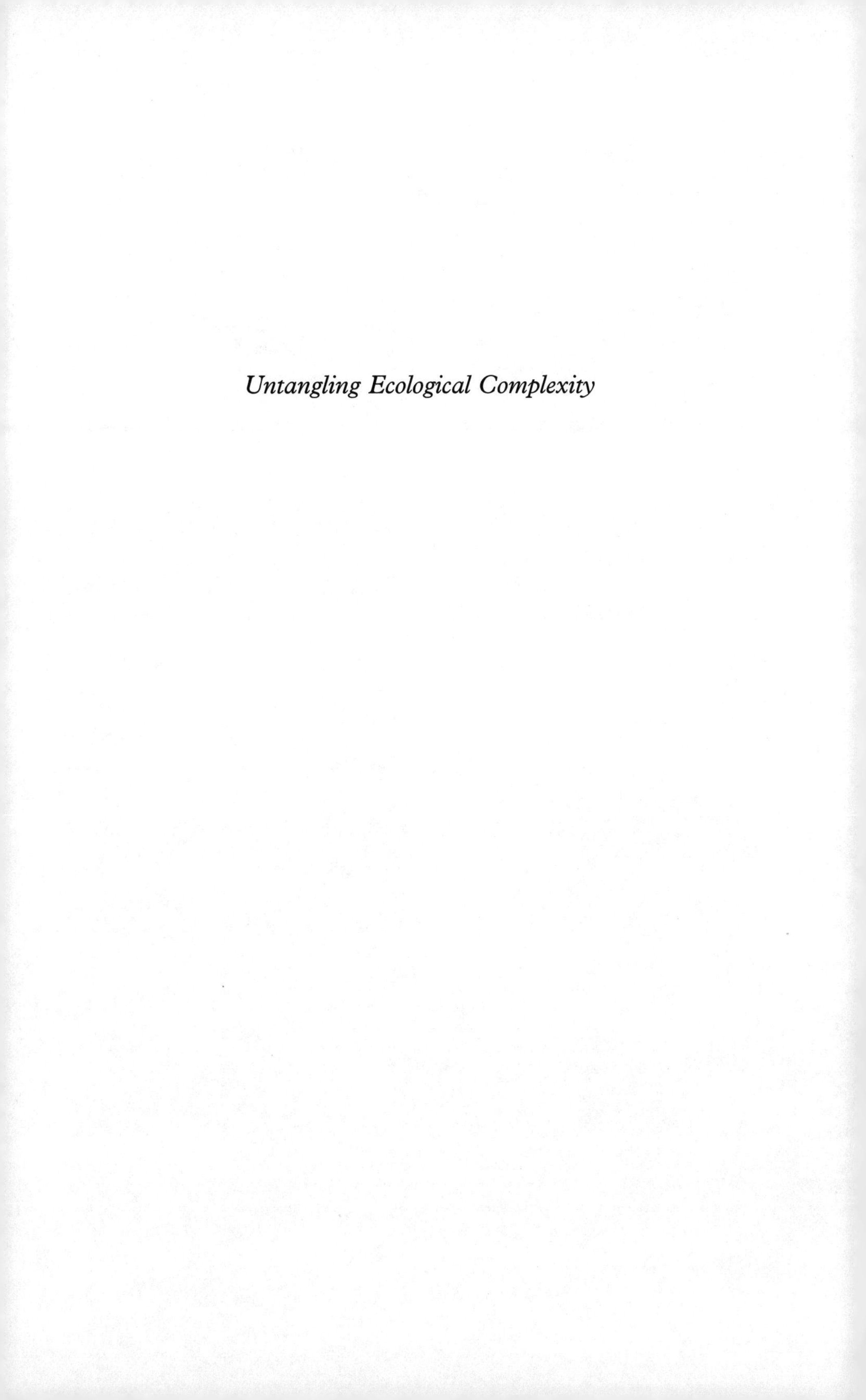

Untangling Ecological Complexity

Untangling Ecological Complexity

The Macroscopic Perspective

Brian A. Maurer

THE
University of Chicago
PRESS
★
Chicago & London

Brian A. Maurer is associate professor of zoology at Brigham Young University. He is the author of *Geographical Population Analysis: Tools for the Analysis of Biodiversity.*

The University of Chicago Press, Chicago 60637
The University of Chicago Press, Ltd., London
© 1999 by The University of Chicago
All rights reserved. Published 1999
08 07 06 05 04 03 02 01 00 99 1 2 3 4 5

ISBN: 0-226-51132-4 (cloth)
ISBN: 0-226-51133-2 (paper)

Library of Congress Cataloging-in-Publication Data

Maurer, Brian A.
Untangling ecological complexity : the macroscopic perspective / Brian A. Maurer.
p. cm.
Includes bibliographical references (p.) and index.
ISBN 0-226-51132-4 (cloth : alk. paper).—ISBN 0-226-51133-2 (pbk. : alk. paper)
1. Biotic communities. I. Title.
QH541.M384 1999
577.8′2—dc21 98-26199
CIP

∞ The paper used in this publication meets the minimum requirements of the American National Standard for Information Sciences—Permanence of Paper for Printed Library Materials, ANSI Z39.48-1992.

Danny, John, and Megan,

this is for your children's grandchildren.

I pray that ideas like these will have made some

difference for their world.

CONTENTS

ACKNOWLEDGMENTS

Any book-length treatment of a subject as complex as community ecology draws upon many sources of information. Although I acknowledge the many individuals with whom I have interacted in various ways while developing the ideas in this book, I accept responsibility for the view set out herein. Any errors are mine, and not those of the people who have helped me. This project was inspired by the year or two that I spent as a postdoctoral associate at the University of Arizona working with Jim Brown. It has taken me many years to capitalize on insights gained during that time. The following persons have read portions of the manuscript or conversed with me on the various subjects discussed in the text. I am indebted to each of them: M. Rosenzweig, M. McKinney, S. Pimm, T. Keitt, M. Taper, P. Marquet, D. Kaufmann, R. Holt, T. Blackburn, N. Pack, E. Linder, E. Schmidt, M.-A. Villard, M. Hopp, R. Lande, M. Lewis, M. Kot, W. Schaffer, M. McPeek, J. Lawton, M.-J. Fortin, S. Gaines, T. Case, Y. Ziv, A. Barnosky, E. Hadly, C. Fowler, D. Jablonski, M. Lomolino, T. Martin, C. Badgley, W. Roberts, R. O'Neill, B. Milne, J. Wiens, J. Rotenberry, T. Root, D. Goodman, G. Russell, P. Nott, J. Curnutt, J. Drake, G. Huxel, J. Kolasa, S. Reeve, D. Montague-Judd, J. Dunning, W. Laurence, D. Simberloff, S. Pickett. I know there are some I have left off this list and I hope that they will be forgiving of my faulty memory. I am especially indebted to the faculty, students, and staff of the Mountain Research Center at Montana State University for the time I spent with them while I was working on the manuscript. The patience and long-suffering of Susan Abrams, and then Christie Henry, throughout the process of producing this book is greatly appreciated. Most of all, I thank Cathy for the love, support, and patience that she has shown while enduring my struggles with this book and everything else.

CHAPTER ONE

Of Entangled Banks and Humble Bees

> When we look at the plants and bushes clothing an entangled bank, we are tempted to attribute their proportional numbers and kinds to what we call chance. But how false a view is this! . . . What a struggle between the several kinds of trees must here have gone on during long centuries, each annually scattering its seeds by the thousand; what war between insect and insect—between insects, snails, and other animals with birds and beasts of prey—all striving to increase, and all feeding on each other or on the trees or their seeds and seedlings, or on the other plants which first clothed the ground and thus checked the growth of the trees! Throw up a handful of feathers, and all must fall to the ground according to definite laws; but how simple is this problem compared to the action and reaction of the innumerable plants and animals which have determined, in the course of centuries, the proportional numbers and kinds of trees now growing on the old Indian ruins!
>
> Darwin (1859)

Of the many problems confronting society, surely one of the gravest is the likelihood that modern human culture may so change the intricate workings of the earth's ecosystems that they will no longer be able to provide the support that human culture requires for its existence. The impact of human culture is astounding in its extent (Turner et al. 1990). Human activities fundamentally alter global processes such as biogeochemical cycles, leading to drastic changes in the functioning of ecosystems important to human activities (Vitousek et al. 1997). Ecological science has an urgent responsibility to help society respond to the changes it has wrought upon the earth. Much of that responsibility lies in providing appropriate guidance about how to preserve biological diversity (Lubchenco et al. 1991). Community ecology, that branch of ecology that deals with associations among species within ecosystems, has much to offer in providing this guidance. Together with population ecology and biogeography, community ecology seeks to understand the mechanisms that cause communities to change. The answers to the questions of community ecology and its related disciplines should provide some of the guidance needed to conserve diversity.

What are the questions that community ecology must answer in order to promote the conservation of biological diversity? Perhaps the most comprehensive statement of the important questions facing community ecology can be found within the Ecological Society of America's Sustainable Biosphere Initiative (SBI; Lubchenco et al. 1991). The SBI prioritized three major areas of research in ecology: global change, biological diversity, and sustainable ecological systems. Within these areas, Lubchenco et al. listed what they perceived as the important intellectual frontiers in ecology and suggested several specific research topics that should guide research at these frontiers. Lubchenco et al.'s list of research priorities and intellectual frontiers involves many questions directly relevant to community ecology. Of these, one of the most fundamental is the documentation of patterns of biological diversity (including species diversity) and determination of their causes. Why would the SBI take such a fundamental problem as identifying patterns of species diversity and claim that it is an intellectual frontier? Have ecologists and their intellectual predecessors not been studying these patterns since before Darwin? Why has the problem not been solved?

The answers to these questions are not to be found in the degree of commitment or creativity of ecologists. Understanding patterns of diversity and their causes remains an intellectual challenge because of the immense effort required to collect and analyze the relevant data and the inherent complexity of causes and effects within ecosystems (Huston 1994). As Huston points out, understanding biological diversity is a hopeless task unless there is some way to break the subject down into manageable pieces. But this raises the question of just what those pieces should be. Some authors have suggested that the pieces exist on different spatial and temporal scales (Huston 1994; Levin 1992; O'Neill et al. 1986; Pimm 1991; Brown 1995; Rosenzweig 1995). This implies that there is not one problem, but many problems interconnected across a range of spatial and temporal scales. The challenge, then, is to develop an approach to community ecology that allows us to conceptualize these connections as rigorously and unambiguously as possible.

This is the task that I address in this book. Ecosystems are complex, but I assume that they are not so complex as to prevent scientists from untangling some of that complexity. As Darwin realized long ago, the complexity observed within the entangled bank cannot be attributed entirely to chance, although there are certainly some elements of uncertainty in the outcomes of events that lead to that complexity. If these outcomes are not entirely due to chance, then there must be regular, deterministic processes that also contribute to what happens in an ecosystem. The challenge of community ecology is to identify that subset of events occurring in an ecosystem that can be attributed to predictable, deterministic processes involving populations of species and their interactions with one another.

A significant proportion of research in community ecology has approached this challenge by attempting to analyze communities through controlled experiments. Where experiments could not be done, a variety of stochastic and deterministic modeling approaches have been tried, most done from the same, reductionist perspective. The reductionist approach to community ecology is in the best tradition of science. It is concerned with establishing unambiguously the factors that determine the nature of species diversity within ecosystems. This emphasis on mechanisms has led to many advances in our understanding of particular communities and of the factors responsible for their persistence within the ecosystems where they are found. But reducing a community to its components via experimental manipulation and modeling cannot identify all mechanisms that are responsible for the persistence and dynamics of a community. There are processes that impinge on communities at much larger scales that need to be considered (Huston 1994). Ecology as a science has rarely addressed the nature of these processes and how they can cause things to happen in communities. This has led several authors to suggest that it is time to expand the spatial and temporal scale of community ecology (Ricklefs 1987; Brown and Maurer 1989; Pimm 1991; Brown 1995).

In attempting to come to some better understanding about what we can and cannot expect to accomplish in this expansion of community ecology, I examine two major questions. The first deals with experimental community ecology. I examine the limitations of this approach with the intent to establish what kinds of processes and mechanisms we are able to study using this approach. Since it is evident that the experimental approach has generated a number of insights into the role of processes such as competition and predation in determining the ability of species to persist together within an ecosystem, it is clear that the experimental approach is a much-needed part of community ecology, and will continue to be so in the future. However, there are questions about other, larger-scale processes that need to be addressed to further our understanding of species diversity (Rosenzweig 1995). It is important to establish which of these questions cannot be answered by the experimental approach, and hence we need to know its limitations.

The second major question that I address is, what kinds of processes operating at larger spatial and temporal scales influence species diversity on the local scales that are commonly studied? A number of such processes exist. Many factors influencing communities at large spatial scales are stochastic from the perspective of the local community. These factors include physical disturbances, such as drought or cold resulting from climate change, and biological ones, such as invasions of competitors, predators, or pathogens. Less often considered are deterministic factors that influence communities at larger scales. For example, population processes such as dispersal can change the outcome of deterministic population processes occurring at smaller scales, such as competition and predation (Holt 1996). Historical outcomes of evolutionary processes may be contingent or random, but the consequences of these processes shape the gene pool of a species, which in turn determines the species' niche. Hence, the ecological interactions among species and their environments occurs within the context of the genetic variation of the species involved. There is some theoretical and empirical evidence that suggests that the niches of species are conservative, in the sense that there is an evolutionary "inertia" that must be overcome in order for new adaptations to evolve (Brooks and McLennan 1991; Ricklefs and Latham 1992; Holt 1996). Given that the outcomes of species' ecological interactions are partly determined by their genetically based adaptations and that these adaptations may often be stable in ecological time, it follows that we might expect to detect some kind of population structure that reflects that stability. In later chapters, I examine the structure of geographic ranges of species and argue that consideration of such structure allows us to develop theories about additional mechanisms that underlie what we observe within local communities. Furthermore, the structure that exists in geographic ranges has implications for our understanding of the operation of mechanisms that determine the regional species pool from which local communities are drawn.

A complete treatment of the nature of these macroscopic patterns in large-scale ecological systems necessarily requires preliminaries. First, it is necessary to establish that statistical processes can indeed lead to macroscopic patterns in complex systems that are not trivial or epiphenomenal. If the macroscopic patterns generated in space and time by complex systems are easily reducible to the action and interaction of a relatively small set of processes operating over short time and small spatial scales, then they hold no additional insights into the nature of the system than patterns that emerge on small scales: the macroscopic patterns can be considered epiphenomena. However, if macroscopic patterns do not reduce in a simple fashion to a reasonably small set of localized processes, it is likely that studying them in their own right will provide additional insight into the nature of the complex system being studied. Second, when considering the nature of macroscopic ecological patterns, it is necessary to understand the nature of the smaller-scale processes that are ultimately involved in their generation. Basic population processes, including those that result from species interactions, may result in different patterns in local communities than they do at regional or continental scales. It is important to get some feeling for what the local-scale patterns are likely to be before venturing a guess at the larger-scale patterns.

Once it is clear that complex systems can generate regular patterns at large scales, and once the nature of smaller-scale ecological processes is evident, then it should be possible to begin developing theoretical insights into large-scale ecological patterns. There are several possible scales at which large-scale patterns might be examined. For example, Robert Holt (1993) aptly described the "mesoscale" as the scale at which populations are aggregated together across landscapes. Aggregations composed of single species are called metapopulations. However, metapopulations do not exist alone, but interact with other species across a range of spatial scales (Caswell and Cohen 1993; Holt 1997). Such "metacommunities" (Holt 1997) are sufficiently large-scale that their analysis reveals many important ecological processes only incompletely examined during the years when Robert MacArthur and his colleagues forged modern community ecology. In addition to the processes of competition, predation, and mutualism that dominated the early development of analytical community ecology, processes that connect different populations together, such as migration, colonization, and extinction, are of interest to ecologists studying this "mesoecology." Mesoecology is a rapidly expanding field, and one that has rightfully attracted a great deal of interest (Hanski and Gilpin 1997).

The scale that I will focus on in this book is the *geographic scale.* The taxonomy of spatial scales in ecology is imprecise, so there is some misunderstanding about what is meant by "geographic" scale. My use of the term in what follows refers to that scale on which entire geographic ranges of species exist. This definition can be problematic when one is considering organisms of very different sizes, but most of what I discuss in later chapters pertains to birds, since they are one of the few groups of organisms that have sufficiently large databases to allow analysis of geographic range patterns. Much of what I discuss must be tempered with the realization that we do not have good enough data on other kinds of organisms to do the kinds of analyses I have done for birds. I hope that large-scale censuses of other groups of organisms will be conducted in the future.

MacArthur (1972) was clearly interested in how patterns of competition and predation resulted in geographic patterns. But there was much work to be done before ecology could grow beyond the borders of the local field studies that have been the backbone of modern community ecology. In an influential paper, Robert Ricklefs (1987) pointed out that the outcomes of many of the processes studied by community ecologists were modified by larger-scale processes occurring at longer temporal and larger spatial scales. This view was reinforced by the empirical and theoretical studies of J. Roughgarden and colleagues (see, e.g., Roughgarden, Gaines, and Pacala 1987), and of many other scientists (see the papers in Ricklefs and Schluter 1993). James Brown and I (Brown and Maurer 1989) argued that not only did these larger-scale processes affect local communities, but they resulted in large-scale patterns that provided new ways to study these large-scale processes. We termed this large-scale perspective *macroecology.* Subsequently, Brown (1995) argued that macroecology was an important synthetic extension of aspects of community and ecosystem ecology, biogeography, and other related disciplines, and should be a field of study in its own right. In the later chapters of this book, I will consider this macroscopic scale in more detail, focusing in particular on the nature of geographic ranges and the implications of macroecology for the evolution of species diversity.

Complexity in Ecological Systems

A conceptual approach to ecology at any scale must deal with an enormous variety of unique events that occur in a myriad of different sequences. The problem for ecologists in dealing with this complexity is how it affects their ability to make generalizations. Ecological science can proceed with relatively few generalizations, because it should theoretically be possible to generate unique hypotheses for each unique event. Yet, as my colleague Bertram Murray pointed out in a talk at a meeting of North American ornithologists, this is not how progress has been made in nonbiological disciplines. According to Murray, physicists tend to be conceptual "lumpers," and develop general explanations from first principles that are used to construct empirical hypotheses. Biologists, on the other hand, tend to conduct empirical studies first, and then develop unique post hoc hypotheses to explain their results. Although there are notable exceptions to this pattern, biological theories often seem to be diverse and fragmented.

The difficulties of developing theories in biology have been a source of much concern among biologists (Mayr 1961; Levins 1966, 1968; Loehle 1983; Peters 1976, 1991; Pickett, Kolasa, and Jones 1994). It is likely that the difference in biologists' and physicists' theoretical styles is due to differences in the subject matter. Since biological systems are thought to be more complex, they are also thought to be less susceptible to generalization than physical systems. Although I believe that there is some substance to this view, I am not convinced that it entirely explains the difference perceived by Murray. In addition, I believe that biologists have been somewhat naive in their approach to biological systems by assuming that the best way to study them is to reduce them to their fundamental units. Clearly there have been rather profound successes in molecular biology, but there are limits to the ability of reductionistic science to explain the nature of ecological systems. This is a direct consequence of their complexity: reduction of ecological systems leads to the bewildering variety of unique phenomena that generated the conceptual problems to which Murray was referring. So Murray's observation is due *both* to the complexity of ecological systems *and* to the naive application of reductionism to them.

This is not to say that reductionist science cannot help scientists understand ecological systems. I am simply arguing that reductionist science *alone* will not suffice. Although physics is often seen as the epitome of reductionism in science, the success of physics is not due to reductionism alone. Nobel laureate Erwin Schrödinger (1944) argued that the precision of physical laws does not derive from the ability to precisely measure the interactions of the fundamental particles of physics. In fact, Schrödinger pointed out that one of the major contributions of quantum physics was to show that there was a tremendous amount of variability and indeterminacy at the microscopic (most reduced) levels in physical systems. This variability exists because it is impossible to measure the velocity and position of individual particles simultaneously. The precision of physical laws arises at much larger scales because these indeterminacies are sufficiently small that they are irrelevant at the macroscopic scale. Laws are the consequence of the interaction of a tremendous number of small particles, each with variable and relatively indeterminate behavior. Schrödinger's point is an important one for ecologists (O'Neill 1989; Powell 1989; Wiens 1989a; Levin 1992). It suggests that if there are any regularities or laws that operate in ecology, they will necessarily be dependent on the scale at which we attempt to measure them. In fact, the distinct possibility exists that there are as yet undiscovered regularities that exist at scales much larger than ecologists have traditionally searched for them. I will return to Schrödinger's argument in the next chapter; suffice it to say here that complexity, per se, is no good reason not to expect the existence of empirical regularities or relatively precise laws.

Theoretical and Empirical Traditions in Community Ecology: The Point of Departure

I want to be very careful to convey to the reader that I am not claiming that laws as precise as those that govern physics exist in ecological systems. I am simply suggesting that we should expect *some* degree of regularity and precision in community ecology, and probably more than has been claimed by others in the past (e.g., Sale 1984; Judson 1994). Furthermore, we should expect that there is a higher likelihood of discovering those regularities if we look at populations and communities at larger spatial and longer temporal scales than has been the tradition in the past.

To undertake a discussion of what some of these regularities are, it is first necessary to get some idea of what community ecology, as a field of study, is. A community can be defined as a group of populations of different species that live together in the same habitat and that affect one another's population dynamics in some fashion. There are several difficulties that arise with definitions like this (Wiens 1989b), and I will discuss some of them later. However, if for the moment we take this as the definition of a community, a few questions arise that can lead to testable hypotheses. The most obvious question is how the population dynamics of one species is related to those of others. A species may increase its population by preying on another; it may remove resources that are necessary for another; or it may in some way enhance the resources available to another. Most studies in community ecology focus on these kinds of interactions. These mechanisms have been tested experimentally both in the field and in the laboratory for specific communities (Diamond 1986; Gilpin, Carpenter, and Pomerantz 1986; Hairston 1989; Underwood 1986, 1997). But changes in populations of species can also be caused by climate, topography, geology, and other physical factors that vary spatially and temporally. Much of community ecology is devoted to sorting out whether for particular communities spatial and temporal dynamics are due primarily to interspecific interactions or to "abiotic" factors such as climate.

The diverse set of ideas encompassed by community ecology can be traced back to at least two different empirical and theoretical traditions. It is beyond the scope of this book to delve too deeply into the development of ideas that now contribute to modern community ecology. Extended treatments are given by McIntosh (1985) and Kingsland (1995). Here, I am interested in the kinds of questions that motivated community ecologists in the past, and how that has led to the modern research program broadly encompassed under the rubric of community ecology.

Complexity and the Beginnings of Community Ecology

In a very real sense, Charles Darwin was the first ecologist. Prior to Darwin, natural history was a discipline devoted to showing the orderliness of nature. Darwin's concept of the struggle for existence, however, cast the interrelationships among organisms and their environment in a very different light. Haeckel coined the term *œcology* to refer to the study of the struggle for existence (McIntosh 1985). Darwin emphasized the complex nature of this struggle (see the quote at the beginning of this chapter), and this complexity was not lost on the early ecologists. As ecology began to emerge as a science, there were two basic responses to the idea of ecological complexity. The first is typified by Stephen Forbes's (1887) discussion of ecological conditions in lakes. The other is represented by Henry Cowles's (1899) description of vegetation on Lake Michigan sand dunes.

Forbes (1887) was impressed with the connectedness of species of plants and animals living together in a lake. Forbes points out that "Nowhere can one see more clearly illustrated what may be called the *sensibility* of such an organic complex, expressed by the fact that whatever affects any species belonging to it, must have its influence of some sort upon the whole assemblage" (537). He goes on to say that because these effects are passed on from one species to the next, it is impossible to study any species in isolation. Hence, according to Forbes, in order to understand the behavior of any single species in an assemblage, we have to understand the behavior of the entire assemblage. Forbes was impressed with how much information was needed to understand the working of any biological assemblage. However, he also recognized yet another level of complexity: "The amount and variety of animal life contained in [lakes] as well as in the streams related to them is extremely variable, depending chiefly on the frequency, extent, and duration of the spring and summer overflows" (538). The complexity inherent in the interactions among individual organisms was also influenced by the seasonal dynamics of water flows, giving rise to "extreme variability" in the distribution and abundance of plant and animal species.

What was the general governing principle that Forbes saw in the diversity of these highly complex lake ecosystems? Forbes believed that in order for interacting species to persist together for any length of time, their net effects on one another had to be positive. He referred to groups of interacting species as having "a close *community of interest.*" The mechanism that caused this community of interest to evolve was natural selection: "Two ideas are thus seen to be sufficient to explain the order evolved from this seeming chaos; the first that of a general community of interests among all of the classes of organic beings here assembled, and the second that of the beneficent power of natural selection which compels such adjustments of the rates of destruction and of multiplication of various species as shall best promote this common interest" (Forbes 1887, 550). Forbes seems to be saying that individuals will find a benefit (in terms of their own fitness) if they do not remove too many resources (or otherwise affect the assemblage in too negative a way), for if they do, the whole system will be less capable of supporting them and their offspring. Thus, in the long run, natural selection should reinforce individual behaviors that contribute to the maintenance of a stable ecosystem. This is different from group selection. In group selection, strictly speaking, it is the group (or community) that benefits from the behaviors, not the individual organism (Wilson 1983). That is, there is a "fitness" of the community that determines its perpetuation relative to other communities. Forbes's "common interest" is similar to modern concepts of adaptive indirect effects (Wilson 1986) and coevolution (Thompson 1994). Forbes argued that community structure is in essence an adaptation of the species living in the community to their mutual interactions and the natural variation in the physical environment. For Forbes, in spite of the tremendous complexity of biological communities, they were structured and stable in the sense that the interactions among species in the community had evolved to allow the species to persist together for long periods of time.

Contrast this view of community structure with that described by Cowles (1899). Cowles studied plant communities on sand dunes on the shores of Lake Michigan. Since the sand dunes on which these communities existed were constantly changing, the composition of the communities was also in constant flux. "Ecology," wrote Cowles, "is a study in dynamics. For its most ready application, plants should be found whose tissues and organs are actually changing at the present time in response to varying conditions. Plant formations should be found which are rapidly passing into other types by reason of a changing condition" (95). The essence of ecology for Cowles was not the coevolved stability of Forbes, but rather the dynamic, ever changing interactions of species with their environments. Ecologists never find repeatable vegetation units across landscapes, according to Cowles; in fact they never find any pair of vegetation units that are identical. Cowles's view differed from Forbes's not in the recognition of complexity, for both of these early ecologists saw complexity as a major problem for ecology. Cowles's view did not allow sufficient time for the organizing force of natural selection to sort out the best set of interactions to stabilize the system. It is interesting to note here that the basic differences in Forbes's and Cowles's views have persisted to the present day: some ecologists see evolutionary dynamics of populations as sufficiently strong and persistent in time and space to allow for the evolution of properties within populations that stabilize the overall community,[1] while others argue that communities are too heterogeneous in space and time to allow the mechanisms of natural selection to weed out variants that tend to destabilize them.

The differences in perspective between Cowles's and Forbes's views of community ecology persisted into the next generation of ecologists. In 1916, Frederick Clements published his book *Plant Succession,* in which he developed Forbes's view to its logical extreme. To Clements, the community was a large organism that was born, developed, and died. Community types evolved into other communities in a way similar to the evolution of species into different species. The climate of a region controlled the nature of the climax, or end stage, of the organic community. The processes that contributed to the "homeostasis" of the climax included biotic interactions among the constituent species; for instance, the most competitive plant species characterized the climax vegetation. One of the important conclusions to be drawn from Clements's ideas was that communities had a natural classification or taxonomy. The basis of this taxonomy was the stable nature of the climate-driven climax (Clements 1936).

A quite different view, following Cowles's ideas regarding ecological communities, was advanced by H. A. Gleason. Gleason (1926) argued that although many vegetation associations could in fact be identified as lasting, stable associations, classifying them into discrete groups was more difficult than previous scholars had implied. According to Gleason, all vegetation associations could be shown to be different in some respect from all other associations, even those that existed in similar places and should fall into the same vegetation class. Often these differences were quantitative, so that species dominant in one association were not dominant in other associations that shared similar species composition. Fur-

1. This view is often translated into the concept of resource partitioning (Schoener 1974). If species within a community have evolved responses to one another that prevent them from using common resources, then the populations of the species involved are able to achieve stable, persistent populations. If this is the case, evolutionary dynamics can be ignored (at least for the short term), and stability can be considered to be related to the specific factors that determine how resources are partitioned.

thermore, Gleason reasoned that many vegetation communities change gradually in both space and time, so that it is difficult to say whether any particular community is one that is the climax condition for that geographic location, or whether it is transitional, in either space or time, between other communities.

There is often some confusion associated with the views of Clements and Gleason that bears mentioning here. First, Clements's view is often associated with an unchanging, static view of community structure. This interpretation most often appears in the discussion of "climax communities," those end points of community succession he envisioned. For example, Odum's (1969) classic paper on ecosystem development postulated that climax communities should be resistant to changes induced by perturbations. Clements, however, wrote of a "dynamic plant ecology," and he was fully aware that many vegetation associations were in flux (McIntosh 1985). His view, however, was that the changes observed in plant communities were explained by (1) the tendency of the environment to set back vegetation conditions to early stages of vegetation succession and (2) the tendency of plant communities to establish interactions among species and with the abiotic environment that lead to deterministic outcomes (i.e., climax communities). Clements invented terms like "disclimax" to indicate that there were some communities in which abiotic disturbances were frequent enough to allow vegetation to exist far from climax conditions for long periods of time. However, he felt that the major force determining the species composition of a community was the tendency of the species to sort themselves according to deterministic outcomes of interactions with each other and with the environment. Second, Gleason's individualistic concept of community structure is often associated with the idea that interactions among species are not important in establishing the composition of a plant or animal association (e.g., James et al. 1984). Gleason realized that there were instances in which species interactions could, given sufficient time, produce large expanses of relatively uniform vegetation.[2] Gleason, however, argued that this was a function of the rate at which environmental changes occurred, so that those areas with the largest degrees of compositional uniformity and stability, at least in temperate climates, were those that had relatively unchanging environments in space and time. Thus, many of the views

2. Gleason's ideas can be reconciled with negative interactions among species, since species populations that randomly occupy a particular site might do so to the disadvantage of other species, and community structure be a case of a "lottery" where species that establish themselves first will determine what the community will be like. It is more difficult to envision how Gleasonian communities might be structured by positive interactions, which are easier to envision as structuring Clementian communities.

of Clements and Gleason were similar. They differed mainly on a single, philosophically profound point. While Clements argued that biotic and abiotic determinants of community structure led to deterministic outcomes, Gleason reasoned that the spatial nature of plant demographics ensured that there would always be a certain level of uncertainty in any community. Ultimately, according to Gleason, and regardless of the kinds of interactions among species in a given community, seed dispersal was sufficiently probabilistic that there must always be some degree of chance involved in which particular set of species was able to establish persistent populations in a given community. Both Clements and Gleason realized that there was a great deal of variation geographically in conditions affecting community structure in plants, but for Gleason, complexity had an additional consequence: the introduction of chance as a causative agent in the development of community structure.

Most practicing community ecologists, whether they study plant or animal communities, would probably argue that Gleason's view is ultimately closer to the truth. In particular, they might cite the unrealistic assumptions required to treat communities, either empirically or theoretically, as organisms. Gleason's view is more satisfying because he was very clear about the kinds of mechanisms that he thought would give rise to plant communities. Because most plants are sessile, plant demography necessarily contains a spatial component: recruitment is profoundly affected by the ability of the offspring of individuals to disperse. Gleason built his individualistic concept of community structure from his understanding of plant dispersal and its effects on migration among different communities; and it seems reasonable to expect that most communities, including animal communities, cannot be considered closed (except perhaps in the unusual ecological conditions found in island habitat).[3] Hence, it seems that Gleason's views are more widely accepted because they are built on more realistic assumptions about communities and seem to be more consistent with empirical evidence about how communities are put together.

Complexity and the Quantitative Description of Communities

Although many community ecologists doubt the existence of the kind of organismal attributes that Clements ascribed to communities, the idea that communities will converge on some kind of stable, repeatable structure as a consequence of biotic interactions has led many to look for ways to describe that structure. Two very general approaches emerged in the

3. But note that even island communities receive a flow of immigrants depending on their size and degree of isolation.

decades after Gleason and Clements that many thought would provide the empirical and theoretical tools necessary to describe and predict the structure of communities. The first was the application of the theory of linear differential equations to communities. This approach was first outlined by Lotka (1920), but soon other scholars realized how this mathematical technique could be used to examine biological communities (Volterra 1926; Gause 1934). Although I will consider this technique in more detail later, a few words about its previous use are in order here, because it was very influential in the theoretical development of community ecology. The second approach was based on the simultaneous description of spatial changes in species abundances of many communities. The statistical technique of ordination was used, especially in plant ecology, to study the covariation of species with each other across environmental gradients (Bray and Curtis 1957; Gauch 1982; Pielou 1984). The term *ordination* implies an ordering of species into groups that are distributed similarly in space and time, and several statistical techniques were developed to facilitate this ordering. One of the more common was based on correlations of species with one another across sampled quadrats or transects (Pielou 1984). The rise of these multivariate statistical techniques in ecology was thought to provide ecologists with powerful empirical tools. I will show in chapter 4 that it is possible, making certain assumptions, to find a relationship between the linear differential equation method and the multivariate statistical method. Here I consider how these techniques influenced the development of community ecology after Clements and Gleason.

Two basic assumptions underlie the application of the theory of linear differential equations to biological communities: (1) that dynamics of species can be modeled as small, deterministic deviations away from their equilibria[4] and (2) that the community is closed to invasion from other species, so that species number remains the same, or possibly decreases, at equilibrium. Note that these assumptions are reasonable under the kind of climax community envisioned by Clements, but are perhaps less so in Gleasonian communities. Given these assumptions, the densities of species when the community ceases to change can be mathematically modeled as linear functions of one another. The major limitation of the linear differential equation method, other than the requirement of restrictive assumptions, is that it is capable of describing a relatively limited array of dynamical behaviors. The densities of species in the community, according to this model, must either (1) converge smoothly, or with

4. This means that species densities are assumed to be very close to their equilibria and that all effects are instantaneous, that is, there are no time lags.

damped oscillations, toward their equilibria; (2) oscillate about their equilibria in fixed periods; or (3) increase without bound smoothly, or fluctuate with oscillations of increasing amplitude, away from their equilibria. The empirical question, and one I shall address in a later chapter, is whether these kinds of behavior are sufficient to describe the dynamics of real communities. It is possible that they only partially explain community dynamics.

Starting with the linear model of community dynamics with its assumptions of equilibrium, a wealth of theory developed between the late 1920s and the late 1970s. This theory focused largely on the role of competition in determining community dynamics, though other kinds of interactions, particularly predation (Hairston, Smith, and Slobodkin 1960; Sih et al. 1985), were examined. Linear community theory was used to reinforce the interpretation that differences between species reduced the degree of competition within communities. This principle, known as the principle of competitive exclusion, was first recognized by naturalists like Grinnell (1917) and Lack (1947), and was later associated with the linear theory of communities by the famous experiments of Gause (1930). McIntosh (1985) pointed out that the idea of competitive exclusion is implicit in Darwin's ideas. Although this is true, quantitative description of the principle, and its use in making predictions about community structure, awaited the work of the early population biologists such as Lotka, Volterra, and Gause.

The work of G. Evelyn Hutchinson (1978) and of his student, Robert H. MacArthur (1972), exemplified the approach of linear community theory and its application to empirical problems. Hutchinson (1958) formalized the concept of the niche as a hypervolume described by the variables that determined the ecological characteristics of individuals within species. He applied the competitive exclusion principle in his definition of the realized niche as that portion of the total possible niche that a species can occupy in a given community of competitors. The implication was that differences in the niche occupied by a species in different communities could be explained by the presence or absence of certain competitors (Diamond 1978). MacArthur (1958) showed that species of wood warblers found in the same habitat used sufficiently different sets of resources so that they could not exclude each other from the community. This idea was expanded by Lack (1971), who interpreted nearly all ecological differences between closely related species as adaptations that allowed them to coexist across geographic space. MacArthur and Levins (1967) termed this concept *limiting similarity,* postulating that species within a community could only share a certain portion of their niche without one species driving the other to extinction. This idea equated interaction coefficients in the Lotka-Volterra equations with the amount of overlap between species' niches.

Many of the ideas of MacArthur and his colleagues fit very well into the general scheme of community structure envisioned by early scholars such as Forbes and Clements. Even if MacArthurian ecology did not require that communities behave as organisms, it argued that interactions among species could lead to coevolutionary adjustments of species such that community structure was enhanced (see treatments in Roughgarden 1979 and Emlen 1984). This expectation led to the concept of assembly rules (Diamond 1975), that is, that the interactions among species could lead to predictable combinations of species across geographic space as a function of the environment in which a particular community was found.[5] Again, although this does not presuppose organismal communities, the idea that community structure is largely deterministic was an integral part of this approach.

A second approach to community ecology began with attempts of ecologists to show that there were statistically recognizable associations of species. In Clements's day, plant ecologists debated whether communities could be described based on subjective assessment of the presence of species or life forms, or whether it was necessary to count individuals of each species in order to quantify the plant community in a given area. As quantification of vegetation increased in importance, it became clear that tables of species abundances still did not provide objective evidence of the existence of discrete vegetation units or associations. To deal with this problem, ecologists began to use sophisticated statistical techniques to describe the abundances of species within communities and the similarities in species composition among communities. The statistical technique of ordination was developed to array species along synthetic statistical gradients that represented groups of species that covaried together (Bray and Curtis 1957; Whittaker 1967). Although several techniques were developed, one of the most widely used is principal components analysis (Pielou 1984). This technique derives these synthetic statistical gradients based on the simple correlations of species with one another.

The quantitative techniques of this statistical approach to community ecology did not presuppose any particular mechanisms that generated community structure (Gauch 1982), rather, the intent was to use them to document patterns of co-occurrence among species. It soon became

5. Assembly rules need not be determined by coevolutionary adjustments of species to one another. Species composition and relative abundance in a community may be determined by an ecological "sorting" process where entry into a community is determined by a species' competitive ability rather than by coevolutionary adjustments.

evident to some researchers that the linear correlations among species used as the basis for many statistical techniques shared a common structure with the linear differential equation approach. Thus, many suggested using multiple regression coefficients to estimate the strength of species interactions with one another (e.g., Hallet and Pimm 1979). Although this method seemed reasonable, it has some significant limitations, which I will discuss in chapter 4.

Community ecology emerged from the 1960s and 1970s as an exciting and rigorous science that had a well-developed body of theory (e.g., Pielou 1977; Roughgarden 1979) and a number of sophisticated statistical techniques that could be used to test that theory (Gauch 1982). There were some initial successes in using this body of theory to predict the outcome of observational (Pulliam 1975; Yeaton 1974) and experimental (Schoener 1982, 1983) studies. However, it soon became evident that empirical patterns in many communities were often different from those predicted by theory. Two general kinds of problems were identified by empiricists. First, patterns in some communities, particularly those in highly variable environments, were too complex to be explained by competition and other biotic interactions (Wiens 1977). Second, patterns in other communities, particularly biogeographic patterns, could often be replicated by simple probabilistic models that assumed nothing more than random associations among species (Connor and Simberloff 1979). By the early 1980s community ecology was embroiled in controversy.

Wiens (1989a) provided a relatively balanced discussion of many of the points of controversy generated by the empirical problems of theoretical community ecology. Even though his book deals primarily with patterns in bird communities, his discussion is wider ranging than the taxonomic bias implies. The point I wish to emphasize is how these empirical difficulties affected the subsequent development of community ecology. There were basically two general types of responses of population and community ecologists to the intense criticisms leveled at them during the late 1970s and early 1980s. First, there was a proliferation of rigorous experiments conducted on local communities.[6] The experimental approach existed prior to the controversies of the last decade, but became a more prominent part of community ecology because it promised to provide a firmer empirical foundation than uncontrolled field studies. Included in this approach was the development of more mechanistic descriptions of interactions among species (Tilman 1982, 1988; Schoener

6. This increase of experimental approaches has not been spread evenly among taxa. For obvious logistical reasons, long-lived or highly vagile species have received less attention than small, short-lived species.

1986), including attempts to understand communities as food webs (Pimm 1982; Schoener 1989). Second, there was an expansion of the sphere of community ecology to include processes occurring on larger spatial and temporal scales (Ricklefs 1987; Ricklefs and Schluter 1993). The goal of this expanded perspective was to account for large-scale processes that affect communities and cannot be manipulated experimentally. A key assumption of this larger-scale approach was that explanations of why a particular community maintains a certain species diversity must go beyond the scale of local manipulations of communities (Huston 1994).

The experimental research program gave rise to elegant experiments designed to tease apart the effects of local physical and biotic factors on populations and communities. This research program has been very successful (Hairston 1989). However, most ecological experiments are done on relatively small spatial and temporal scales (Baskin 1997). This limits the extent to which the results of these experiments can be generalized or applied to other systems. Since we know processes occurring at larger spatial and temporal scales influence what happens in local communities (Ricklefs 1987), we must admit that what we can learn about communities using small-scale studies will be limited (Brown 1995). I discuss some of these limitations in more detail in chapter 3, but I present here a hypothetical example to illustrate the point. Suppose that a particular set of ecological experiments shows that at one site, a species is superior to a similar competitor. Furthermore, suppose that the experiments show that the superior species uses resources more efficiently, and can therefore deprive the competitor of a significant share of them. There are at least two major limitations to the inferences that can be drawn from this experiment. First, the results of the experiment cannot be generalized to other communities. Just because one species is a superior competitor in one community does not mean it will be superior in a community in a different place with different combinations of ecological and physical conditions. Conditions may exist in different communities that cause a species that is inferior in one place to be superior in another (Dunson and Travis 1991). Second, the experimental approach cannot provide an explanation of why this might be so, except to say that competitive relationships change from one place to the next. This is one reason that many experimental ecologists have doubted the existence of general laws in ecology.

The expanded scale of the macroscopic approach to community ecology cannot provide the controlled experimental rigor of studies done on smaller systems. We cannot replicate biotas and continents to determine the effects of geographic factors on local community structure. But the expanded scale of this newer approach to species diversity has led to intriguing discoveries of patterns that appear to be very general (Brown 1995).[7] Even more encouraging is that sometimes these patterns lend themselves to elegant explanations that generate hypotheses that can be tested with independent data sets. Thus, the larger-scale, macroecological approach has already provided the kinds of generalizations lacking in more spatially and temporally restricted studies.

This apparent emergence of generalities from macroecological studies begs for some explanation. A continental biota, such as the mammals of North America, is an exceedingly complex system and its definition is arbitrary. Indeed, it is arguable that such an assemblage is really composed of many complex systems (such as populations, ecosystems, and gene pools) that interact with one another in many ways and at many spatial and temporal scales. A mechanistic analysis of such a system is beyond our powers of description. Why then would we see regularities among geographic-scale assemblages when we rarely see them among local communities? This paradox I leave to the reader to contemplate for the time being.

A Brief Look Forward

The goal of this book is to argue that in some systems, regularities—"simplifications" of a sort—arise because of constraints that are inherent in the system and its environment. Such regularities will be most evident in large-scale systems because the idiosyncrasies of individual components can no longer obscure constraints imposed on the system from within and without. Assemblages of species on continents are an example of such constrained complexity. When the constraints operating on a complex system are sufficiently strong, they impose an ordering on the behavior of the components. In what follows I will consider some aspects of this tendency of complex systems to become ordered by internal and external constraints (chap. 2). I then examine how this different perspective suggests that we view local communities (chap. 3). There are two alternative models for complex communities that are considered in the following two chapters: the linear community (chap. 4) and the nonlinear community (chap. 5). Each of these perspectives has appeal, but neither is entirely satisfactory. The choice between them might have something

7. One example is the observation that in all compilations of the body sizes of groups of related species on continents examined thus far, the statistical distribution of log body masses among species is always positively skewed (Bonner 1988; Maurer, Brown, and Rusler 1992).

to do with the spatial or temporal scale on which one chooses to study a community. In chapter 6 I examine empirical patterns that form the core of the emerging field of macroecology. I consider the idea of geographic range structure in chapter 7, with the goal of determining if the general patterns that emerge from examining geographic ranges can give rise to general principles. In chapters 8 and 9, I examine applications of some of the insights derived from macroecology. If there are regularities at the geographic scale, this might lead to new ways of viewing local community patterns (chap. 8). Furthermore, the patterns that arise from macroecology have very definite implications for the evolution of species diversity (chap. 9).

CHAPTER TWO

From Micro to Macro and Back Again

> It so happens that many of the components that play an important role in nature, both organic and inorganic, are built up of large numbers of individuals, themselves very small as compared with the aggregations they form. Accordingly, the study of systems of this kind can be taken up in two separate aspects, namely, first with attention centered upon the phenomena displayed by the component aggregates in bulk; we may speak of this as the *Bulk Mechanics* or *Macro-Mechanics* of the evolving system. And secondly, the study of such systems may be conducted with the attention centered primarily upon the phenomena displayed by the individuals of which the aggregates are composed. This branch of the subject may suitably be termed the *Micro-Mechanics* of the evolving system. It is evident that between these two branches or aspects of the general discipline there is an inherent relation, arising from the fact that the bulk effects observed are the nature of a statistical manifestation or resultant of the detail working of the micro-individuals. The study of this inherent connection is, accordingly, the special concern of a separate branch which we may speak of as *Statistical Mechanics* . . . but . . . its scope shall be extended so as to include the statistical treatment of the dynamical problems presented by aggregates of living organisms.
>
> A. J. Lotka (1925)

Without question, one of the most remarkable properties of the systems studied by ecologists and evolutionary biologists is their inherent variability. Observed variability is of sufficient magnitude that even though biological systems show regularities, no two biological systems are identical. The application of this truism ranges from the properties of individual cells to those of biomes. In the face of such variability, the interesting question is, how do regularities in the structure of biological systems become and remain stable? I suggest below that regularities in biological systems arise through a statistical process of causality. That is, biological systems that are composed of a sufficiently large number of smaller entities (such as a cell composed of many interacting molecules) are able to develop stable structures because although the effects of the relationships between individual components are small, when added together they determine the fate of the entire system. This does not preclude any one subset of these entities from having a profound effect on the system, but that effect is due to such entities' altering the interactions among many other entities in the system. Structures arise in such complex systems when components react to constraints imposed by the history of changes that the system has undergone. Thus, the system behaves in a cohesive manner. In this chapter I explore insights about structure that arise from treating complex systems as nonequilibrial, statistical systems (e.g., Brooks and Wiley 1988; Laszlo 1987; Kauffman 1993). I then examine how the application of these ideas regarding nonequilibrial systems has provided insights into the nature of complex biological systems.

I begin by examining the relationship between variability, uncertainty, and structure. Although this relationship has been dealt with extensively by many authors (e.g., Conrad 1983; Prigogine and Stengers 1984; Brooks and Wiley 1988), the general concepts are relatively straightforward and I present them to establish a definition of a statistical system. Next, I examine the structure of statistical systems and show that the explanation of structure in such systems depends on (1) an understanding of how the behavior of individual entities that make up the system determines the kinetics of the system itself and (2) an understanding of the constraints on a system's kinetics imposed by its participation as an entity in a larger system. I then review statistical thinking in biology from both empirical and theoretical viewpoints. I show that the application of statistical thinking in data analysis has been very different from its application in biological theory. Next, I develop a general form of statistical paradigms used to describe biological systems. Finally, I examine how statistical paradigms can be made empirically operational in order to suggest how the generation and testing of empirical hypotheses regarding complexity in biological systems can further our understanding of such systems.

Concepts of Variability and Uncertainty

Stochasticity

Much biological thinking appeals to "stochasticity" as a factor that influences the nature of biological systems. However, as Strong (1983) pointed out, such appeals are often sufficiently imprecise so as to introduce "unnecessary metaphysical problems" into discussions of the properties of biological systems. Often the concept of stochasticity connotes an inability to determine causality. Thus, for example, Wiens (1984) suggested that the less a community is structured by competition and other biotic processes, the more it is structured by stochastic factors.

The inability to determine causality ultimately arises from a lack of information about the system being considered. Lack of information is in turn a consequence of (1) the large number of entities contained in the system, (2) the variability among those entities, and (3) uncertainty about the behavior of the entities.[8] Because of the large number of entities and their variability, the system is too complex for the human mind to describe. Only processes that affect most of the entities in the system in the same way so that they respond in a sufficiently similar manner will be amenable to cause and effect analysis. Other processes that may be influencing different parts of the system in different ways will be difficult to analyze. It is convenient, for the purposes of describing the system, to represent the effects of such processes using probability models of sufficient generality and simplicity. Thus, the description of most ecological systems can be accomplished by using a model partitioned into causal mechanisms and random, or stochastic, effects. However, it is often the case that both the mechanistic and stochastic components of such models are imprecise representations of the true system.

Schrödinger's Vision of Biological Structure

On first reflection, it might seem that complex systems composed of many, variable entities should by definition not exhibit any structure, because structure implies regularities, and a complex system is complex because such regularities are not obvious. In considering the nature of life, Schrödinger viewed this problem from his standpoint as a physicist and one of the founders of quantum mechanics. One of the more intriguing ideas Schrödinger discussed was the concept that "the laws of physics and chemistry are statistical throughout" (1944, 4). A basic assumption of quantum mechanics is that matter is ultimately made up of a large number of small particles with stochastic properties. Thus, in his view, regular, repeatable laws of classical physics were only approximations. The reason they seem so regular to us is that there are so many small particles involved in phenomena like gravitation, magnetism, and diffusion, that the uncertainty they encompass becomes insignificant.[9]

8. In quantum mechanics, the entities that make up the systems being studied are inherently indeterminate. In principle, larger systems like organisms should be perfectly predictable if they obey the laws of physics, but in practice it is impossible to obtain and process enough information about them to enable one to perfectly predict their behavior.

9. The uncertainty is on the order of $n^{-0.5}$, where n is the number of particles involved. Thus if, in a given physical system, there are 1,000 particles, then approximately 3% of the time we would make errors in describing the nature of the system, which is pretty low precision for a physical law! If there were 1 million particles, the error would be about 0.1%, and so on. For a liter of gas molecules, the error in describing the physical properties of the ensemble would be about 10^{-17}%, a negligible probability indeed!

This law of large numbers hides, so to speak, the uncertainty of these physical processes from us, and it is only when physicists were able to detect and study these small particles that the uncertainty became apparent.

So at the microscopic level, even in the most precise of physical systems, there is a tremendous amount of indeterminacy. This indeterminacy is resolved at the macroscopic level, however, by the accumulation of a sufficient number of particles, each of which has only a small effect on the ensemble, so that the emergence of precise natural laws is virtually a certainty. The question, then, arises as to whether the processes governing biological systems are fundamentally different from those governing physical systems. Although biological systems are typically composed of fewer components (by many orders of magnitude), is it possible that they might form statistical ensembles that behave in a relatively precise, lawlike fashion? Schrödinger's answer to this question is surprising. He cleverly deflects the reader's attention by noting that "from all we have learnt [*sic*] about the structure of living matter, we must be prepared to find it working in a manner that cannot be reduced to the ordinary laws of physics" (1944, 76). Then he shows that it is the laws of physics that themselves are approximate and, consequently, no different in kind from biological phenomena! What causes the regularity in physical systems made up of variable and indeterminate particles is that constraints are imposed upon them by their nature as solids, "which are kept in shape by London-Heitler forces, strong enough to elude the disorderly tendency of heat motion at ordinary temperature" (85). Similarly, biological systems have order imposed on them because they use solar energy (or its derivatives) to maintain a large number of relatively indeterminate particles (e.g., proteins, cells, etc.) within strong physical constraints.

Do these considerations apply to communities and ecosystems? Strong constraints seem to exist within an organism so that it behaves as a single ensemble, but communities and ecosystems are not like organisms at all. They have inherently fewer constraints, and behave more like loosely connected systems composed of a relatively small number of interacting entities. Communities, at least at the small scales at which they are commonly studied, are more like parts of larger systems than like cohesive wholes. This realization has led to the emergence of landscape ecology (Forman and Godron 1986; Urban, O'Neill, and Shugart 1987; Kotliar and Wiens 1990), where the major focus of study is movements of organisms, nutrients, and energy between communities within larger landscapes. Communities, then, are probably the wrong kinds of entities to look at for high degrees of organization.

The Relationship between Variability and Structure

Since Schrödinger, many biologists and physicists have worked on the problem of how regularities might arise in biological systems. There is no question that biological systems have such regularities: All genes are made of nucleic acids. All living things, perhaps with the exception of viruses, if one considers them living, are made up of one or more cells. New organisms are built from single cells through a regular sequence of developmental events. When sexual reproduction exists, it usually has rather profound evolutionary consequences. Virtually all adaptations result from some type of natural selection. Clearly, these few examples suggest that biological nature, although complex, is certainly not without general governing principles. In this section I examine the question of how such general principles might arise in complex systems.

In a system composed of many, variable entities, the dynamics resulting from interactions among those entities are exceedingly complex. In fact, Prigogine and Stengers (1984) contended that the dynamics of such systems are so complex that precise predictions of changes in the system based on its initial conditions are impossible.[10] In such a system, as the time between initial conditions and the target prediction point increases, so does the uncertainty of the prediction. Thus, variability and complexity lead to indeterminacy: small changes in initial conditions may lead to divergent end results as the system changes over time.

Some temporal sequences that a complex system can undergo will lead to a dissolution of the interactions among its many components. As an example, think of the effect that cancer has on individual organisms. Only systems that have low probabilities of following destabilizing trajectories are likely to develop or maintain structure. One of the major difficulties faced by scientists attempting to reconcile biology and classical thermodynamics has to do with the ability of biological systems to "choose" trajectories that maintain highly nonrandom interactions among their components. Equilibrium thermodynamics postulates that most changes in systems lead them along trajectories that decrease the order in interactions among system components, as a consequence of the second law of

10. This is a property that some scientists have suggested be used as a general definition of a chaotic system (e.g., Ellner and Turchin 1995), although I believe that the label "chaos" for these systems is a misnomer (see the next section). Generally, in chaotic systems, the temporal dynamics of the system are governed by sets of nonlinear equations. It is these nonlinearities that describe the complexity of the system, and they can lead to very complex dynamics. One of the properties of these complex dynamics is that the ability to predict the future course of the dynamics of the system declines exponentially with time.

thermodynamics.[11] Hence, according to an equilibrium thermodynamic view, biological systems are highly unlikely.

A resolution to the problem posed by equilibrium thermodynamics has been suggested by a number of scholars. In cosmology, for example, models of the expanding universe suggest that the possible number of states that the mass of the universe can occupy increases much faster than the states that it actually does occupy (Frautschi 1988). This results from a kind of balance or tension between the tendency of matter to lose useful energy as a consequence of the second law, and the tendency of matter to attract itself due to gravitation. One way to think of this is that the thermodynamic equilibrium of the universe "pulls" matter toward greater disorder, while the gravitational equilibrium of the universe has the opposite effect. This tension between gravitational and thermodynamic equilibrium in the universe allows the spontaneous formation of nonrandom aggregations of matter (e.g., stars, solar systems, etc.). Thus, gravity acts as a constraint on the possible evolution of the universe. The energy produced by the "big bang" at the beginning of the universe can then be "channeled," if you will, into interactions among the units of matter in the universe so that local order increases, regardless of the ultimate fate of that matter.

The expanding universe model can be generalized to apply to other complex systems. If there are constraints that impose some degree of cohesion on the components of the system, energy inputs into the system can be channeled so that possible dynamical trajectories leading to dissolution of the system have low probabilities of being realized (Collier 1988). The role of constraints that impose cohesion on the system is very important. There are several factors that can constrain system components to interact with one another. System components may have inherent properties that give them a natural tendency to interact. For example, in a population of sexual, outcrossing organisms, individuals interact with one another via reproduction. System components may also be constrained to interact by physical barriers or gradients. On islands, for example, physical barriers limit the diffusion of populations beyond the island, and hence force individuals in the population to interact.

We have established so far that systems with a large number of components have indeterminate temporal dynamics and will break into their component parts unless there are constraints imposed on the system to

11. The second law of thermodynamics can be stated in a number of forms. Here we will use the following definition: any transfer of energy between system components is accompanied by an energy loss as heat. Thus, the second law requires that interactions among components of a system become less ordered over time.

maintain interactions among parts. The result of coupling system indeterminacy with cohesion among parts is that system dynamics become irreversible. Once the system undergoes any set of changes, it cannot return to exactly the same configuration from which it started. Prigogine and Stengers (1984) referred to this as the "entropy barrier" that a system cannot transgress. This irreversibility has important consequences. First, it means that even if the system appears to be similar to its condition prior to a given set of changes, the probability of it being identical is vanishingly small. Second, irreversibility leads to constraints on the future changes that the system can undergo. Because of the past dynamics of the system, which have determined its current configuration, some changes are not possible. Thus, in addition to constraints imposed by component interactions and the external environment, complex systems can be constrained by their history of changes (Brooks and Wiley 1988).

A system that is historically constrained to a limited set of configurations and dynamics often exists in competition with other systems seeking to use the same energy source. Systems that capture the most energy will displace other systems if usable energy is in limited supply (Lotka 1922a,b; Odum 1994). Most of the time, the efficiency and energy-capturing ability of a system depends on the reliability and efficiency of the interactions among its parts. Thus, on the average, within a population of competing systems the best-integrated systems will be favored over time. Biologists should recognize that this is an extension of the principle of natural selection.[12]

So what I have argued is that competition for energy among complex systems will favor those systems with greater energy-processing capabilities, which often, though not always, are enhanced by the efficiency, reliability, and stability of interactions among components. In particular, when there is a source of free energy available to a population of complex systems, and persistence depends on capturing adequate energy to maintain and perpetuate the system, on the average systems that are more organized will persist at the expense of those that are less organized. If this is true, then over time there should be a tendency for the most stable, or structured, systems to accumulate. By structure, I mean that the entities within the system interact or relate to one another in a consistent and stable manner so that the overall persistence of the system is enhanced.

It should be noted that there is an additional property of living systems that makes them more likely to increase in structure and complexity over

12. One of A. J. Lotka's unheralded contributions was the recognition that natural selection could be stated as an energetic principle (1922a,b), as I have done here.

time than other complex systems. Living systems are highly constrained; in fact, these constraints are physically encoded into their structure in the form of nucleic acids. Compare this with a geomorphological system such as a mountain range composed of many different types of rocks. Such a system is composed of many atoms bound together by strong cohesive forces into large, complex crystals. These crystals undergo complex dynamics that are irreversible: it is impossible to recreate a mountain once it has eroded. Gravity and weathering tend to break down the mineral crystals in a mountain and transform them into a relatively unstructured, low-energy state such as an alluvial plain. The differences between such a complex system and a living system are twofold. First, a living system actively acquires resources and uses them to generate internal sources of energy used to do biological work. Geomorphological systems are acted upon by external sources of energy that provide chemical and mechanical energy to break up complex mineral structures into simpler ones. Second, living systems contain information about their own history in the form of genetic information. This information is used to constrain the mode of energy acquisition and processing of the system. Essentially, a biological system "remembers" what worked in the past, and "tries out" various combinations of genetic information to "test" potential "solutions" to environmental problems. In a geomorphological system there are no historical constraints, because any information about the environment experienced by such a system is lost when its structure is changed by external forces. The constraints that determine the state of a geomorphological system are purely physical.

To sum up the argument to this point, we can postulate that the ultimate result of complexity in dynamic systems that are able to store information about their environment is the generation or maintenance of structure. In most systems studied by biologists, the number of components contained in the system is fairly large (although not as large as in many physical systems). Describing how each individual component contributes to the overall structure of the system is for the most part a practical impossibility. However, if each component contributes a small effect to the overall structure of the system, then it should be possible to describe the system in statistical terms by relating the probability distribution of regularities in the properties of components to the properties of the system. If such a description is possible, then the system is a statistical system. Not all biological systems are amenable to such description, but many are. In what follows, I examine the usefulness of statistical descriptions and suggest that there are many ecological and evolutionary systems that can be described as statistical systems.

Hierarchical Structure in Biological Systems

Modeling Kinetics in Hierarchical Systems

Complex systems that are capable of information storage and retrieval, then, if provided with sufficient energy and appropriate constraints, should maintain complex structure. In biological systems, and perhaps in other complex systems as well, the structures that are often achieved are hierarchical in nature (Allen and Starr 1982; Eldredge and Salthe 1984; Eldredge 1985; Salthe 1985; O'Neill et al. 1986). How do the mechanisms described in the previous section operate in a system that is hierarchically organized to maintain structure? Salthe (1985) argued that in order to understand the kinetics of any given level in a biological hierarchy, call it the focal level, it is necessary to examine the properties of the next lower level in the hierarchy as well as the constraints imposed on the focal level by its being a part of the next higher level. Interestingly, Salthe argued that relationships among levels are nontransitive. A consequence of this is that in order to understand the focal level, one does not need to go beyond the adjacent levels for explanation. It is also significant that in Salthe's model the focal-level kinetics are determined statistically by the properties of individual entities in much the same way as I have argued above.

A simple example illustrates how Salthe's model works. Suppose we choose as the focal level a large population of asexually reproducing organisms. Our aim is to describe the population kinetics. We know that changes in population density result from births and deaths of individual organisms, but the birth or death of any single organism will have little influence on population density. What then is the connection between the lower level, organisms, and the focal level, kinetics? Furthermore, if we let the next higher level be that of the ecosystem, how does the ecosystem influence population kinetics? Much of the theory of population ecology is centered on models which describe the per capita rate of growth. In the simplest formulation, the per capita rate of growth is simply the difference between the birth (*b*) and death (*d*) rates. In a population that grows in discrete time this can be written

$$\frac{N(t + \Delta t)}{N(t)} = b - d, \tag{2.1}$$

where Δt is the time period during which population change occurs. Note that, in this formulation, all organisms in the population are assumed to be identical. This is clearly unsatisfactory, because there is no question that individual organisms make different contributions to population dynamics. Hence, this simple model of population dynamics makes unrealistic assumptions about the nature of individual organisms. It obscures the nature of the relationship between the population level and the organismal level.

In order to clarify the relationship between the organismal level and the population level, assume for a moment that the ecosystem level is constant and unchanging, and thus has a fixed effect on lower-level processes that is independent of both time and the number of organisms in the population. Now suppose that each individual in the population has probabilities of giving birth to a new individual (λ_i) and of dying (μ_i) that are constant from one time period to the next and are independent of population density. The total number of births in the interval $t + \Delta t$ can be represented by the random variable $B = \sum X_i$, where the random variable X_i equals one with probability λ_i and equals zero with probability $1 - \lambda_i$. Likewise, the number of deaths in the interval is also a random variable $D = \sum Y_i$, where the random variable Y_i equals one with probability μ_i and equals zero with probability $1 - \mu_i$. Assuming for the moment that the number in the population at time t is fixed, the random variable $R = (B - D)/N(t)$ is the per capita rate of growth over the finite interval of time $t + \Delta t$. The expected value of the random variable R is

$$\begin{aligned} E[R] &= \frac{\Sigma\lambda_i}{N(t)} - \frac{\Sigma\mu_i}{N(t)} \\ &= \lambda - \mu, \end{aligned}$$

where λ and μ are average probabilities of birth and death, respectively. Thus, the expected value of the random variable R is the mean of a stochastic birth-death process, and is mathematically similar to the deterministic version given above in equation (2.1). The important point to note is that the birth and death rates are averages determined by the probabilities of birth and death of individual organisms. Thus, in this simple formulation, the connection between the focal (population) level and the lower (organismal) level is statistical (Salthe 1985).

In order for the population to behave as a unit rather than as a collection of individual organisms, the organisms must behave in some cohesive fashion. Cohesion in a population can come from two sources: (1) It can result from interactions among individual organisms. For example, if each organism required a certain amount of space to exist and reproduce, and its share of space was affected by other organisms, then direct interactions among organisms would cause changes in the probabilities of birth and death as a function of the number of organisms in the population. (2) It can be forced upon the population by the ecosystem. The ecosystem provides energy and nutrients for each individual to use for maintenance and reproduction. If the removal of resources by one organism affects those available to others, then the patterns of resource renewal in the ecosystem can have a profound influence on the probabilities of survival and reproduction of individual organisms. Individuals therefore interact indirectly through shared biotic resources. Lomnicki (1988) has discussed some simple models of population growth that indicate that asymmetrical interactions among organisms can lead to population models of greater stability than models without such asymmetry.

From the preceding discussion, two important concepts emerge about the mechanisms generating regularities in hierarchically organized systems like biological populations. First, the properties of the focal level (e.g., population) are determined in a statistical sense by the objects that are contained in the next lower level (e.g., organismal). Second, cohesion among those objects is a consequence of their being constrained to a subset of possibilities by two factors: (1) their interactions with other objects at the same level and (2) external conditions forced upon them by processes occurring at higher levels.

There may be certain patterns of interaction among objects at the lower level that increase the likelihood of stability at the focal level, as Lomnicki (1988) has shown for some population models. Thus, in addition to statistical processes and cohesion, the history of interactions among lower-level objects may determine the current composition of the focal level. Thus, historical constraint can also contribute to hierarchical structure. The role of historical constraint in hierarchical systems has not yet been examined from a mechanistic viewpoint, although several authors have provided heuristic discussions of the problem (Eldredge 1985; Salthe 1985; Brooks and Wiley 1988). In biological systems, the mechanism of historical constraint is known. As noted above, historical information is retained in the transcribed regions of DNA, which in turn applies that information as constraints on the physiological, morphological, and behavioral development of individual organisms. Individual organisms, in turn, participate in population processes, interactions with organisms of different species, and as parts of energy- and nutrient-processing components of ecosystems. Thus, any ecological phenomenon that includes the participation of individual organisms will be constrained by the historical information that constrains the properties of organisms. There is much more that needs to be learned about these connections between populations, communities, and ecosystems, and the genetic information in the gene pools that constrains the properties of organisms.

Unlearning the "Lessons of Chaos for Ecology"

Complexity in some systems can be represented by deterministic equations that have nonlinear terms in them. The first applications of such equations to ecological populations indicated that very simple nonlinear equations could generate fluctuations in abundance that are qualitatively similar to the kinds of dynamics seen in many natural populations (e.g., May 1974, 1976; May and Oster 1976). This created a great deal of discomfort among some ecologists, because it meant that the dynamics of any ecological system governed by nonlinear dynamics could not be predicted with any certainty (Hastings et al. 1993; Judson 1994). If the fate of ecological systems cannot be predicted, then this must mean that there are no general laws that define ecological systems: each system must be approached from a unique perspective. "Thus," one reviewer concluded, "the lessons of chaos are that simple principles will be the exception, incidents of history are important, and that, because predictions are impossible, the only way to find out what is going on in some dynamic systems is to simulate them exactly" (Judson 1994, 11).

This is a rather gloomy view of ecology. If nonlinearities predominate in ecological systems, as Robert May predicted (1976, 467), then according to the gloomy view of ecology, we should not expect there to be any regularities, commonalities, general principles, or anything like them, except in a few peculiar situations. I take exception to this view, because I think that there are quite a few generalities. Natural selection, for example, is a pervasive, general principle that applies to virtually every biological system at one level or another. And natural selection is based on the Malthusian principle: that populations will increase as rapidly as possible until constrained by their environment. These two principles are very familiar to ecologists, and I am convinced that ecologists have come up with others, such as competitive exclusion. As we shall see later, one of the major achievements of modern population genetics has been to show that natural selection can only be understood as a statistical phenomenon.

The lessons of chaos that many ecologists seem to have learned are, I believe, somewhat misguided (for a similar view, see Hastings et al. 1993). The emphasis in the past has been reductionist: systems have been studied by examining the individual sequence of values that the state variables (e.g., population abundance) take on. Because forecasting changes in the state of a chaotic system becomes rapidly less reliable over time, the assumption is that a chaotic system has no structure. But the reality is that most of the "canonical" chaotic systems studied exhibit rather striking regularities. The regularities can only be seen at larger scales. They appear in the statistical distributions of values that the state variables can take on.

Let me give a simple example to illustrate this important point. The logistic equation of population growth has been widely applied to biological populations. May (1975) showed that various discrete time analogues of the logistic equation could, under certain conditions, generate chaotic dynamics. Consider the following equation, referred to as the Ricker equation:

$$N(t + \Delta t) = N(t) \exp\left[r\left(1 - \frac{N(t)}{K}\right)\right]. \qquad \textbf{(2.2)}$$

The nonlinearity in this equation comes from the exponential term. This is one way to write the logistic equation in discrete form (May 1975). The behavior of equation (2.2) is determined by the growth rate parameter r. When r exceeds a value of about 2.69, the resulting dynamics generated are chaotic. An example of this kind of chaotic behavior is given in figure 2.1A. Note that though both time series in the figure were generated with identical parameter values ($r = 3.7$, $K = 50$), and had nearly identical initial values, after a short period of time the time series look very different. But these differences belie the true similarity between the time series, because if the statistical distributions of values achieved by these two time series are compared, they are nearly identical (fig. 2.1B). In fact, I iterated this process 500 times and calculated standard deviations for the frequency distributions like those in figure 2.1B. Coefficients of variation were typically around 10% for time series of 1,000 time steps (fig. 2.2).

Some chaotic time series generated from equations like equation (2.2) are stable in a statistical sense. That is, the statistical distributions that describe the values that the time series visits are themselves unchanging. Furthermore, Holt (1983) showed that the time average of $N(t)$ is simply K, so the mean of all such time series will be identical. The fact that time series begun with different values using equation (2.2) will converge on the same statistical distribution of abundances is a characteristic of some chaotic time series technically described by the property of *ergodicity* (Ruelle 1989; Lasota and Mackey 1994). An ergodic chaotic system will have every one of the solutions for a particular set of parameters describe the same probability distribution. Any two time series that are generated by an ergodic chaotic system but that have different probability distributions will be associated with different parameter values (fig. 2.2). Thus, from a statistical viewpoint, ergodic chaotic systems are, in fact, stable, even if the individual values that the time series generates are unpredictable.

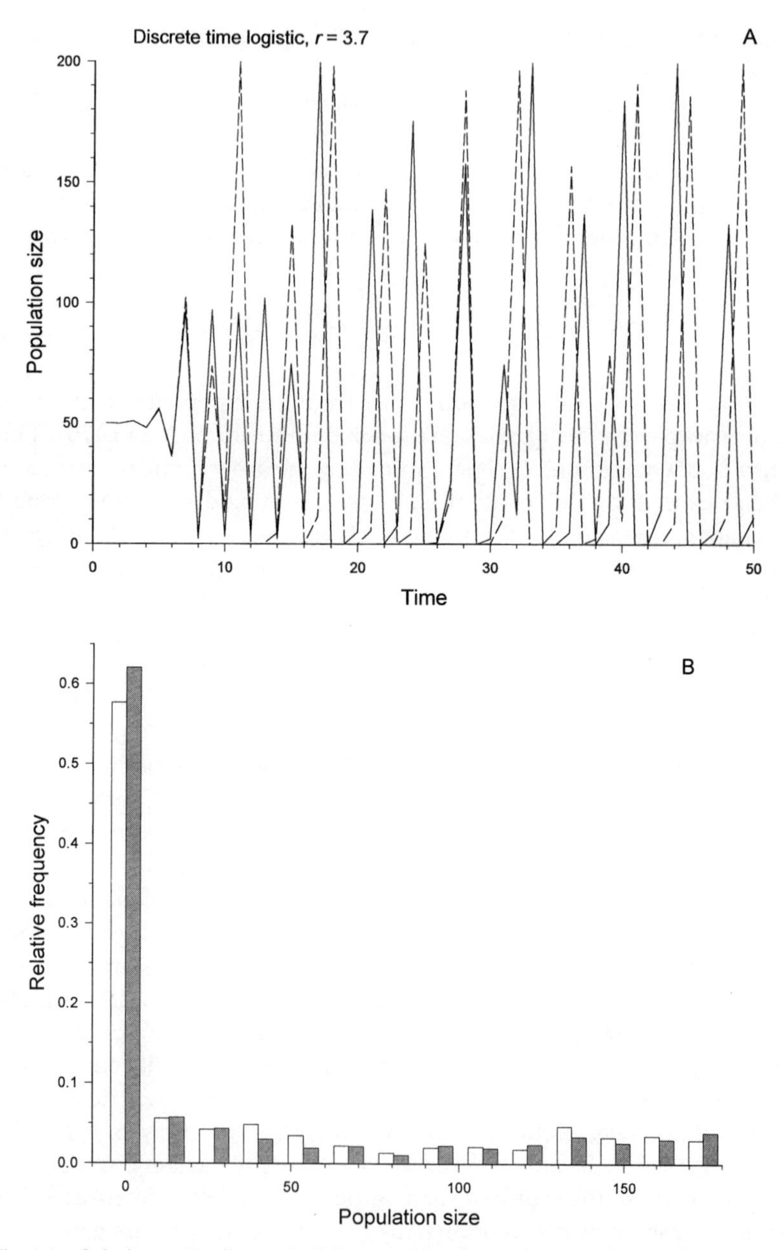

Fig. 2.1 Solution to the discrete logistic equation of population growth when $r = 3.7$ and $K = 50$. *A,* time series for two different starting times that vary by a tiny amount (one time series was started with 50.1 individuals, the second with 50.11 individuals). Note that the time series are nearly identical for the first few iterations, then get quite different as time goes on. The sequence of values shown here is only for the first 50 points in the time series. *B,* relative frequency histograms for 1,000 iterations of equation (2.2) for the two time series shown in part in panel A.

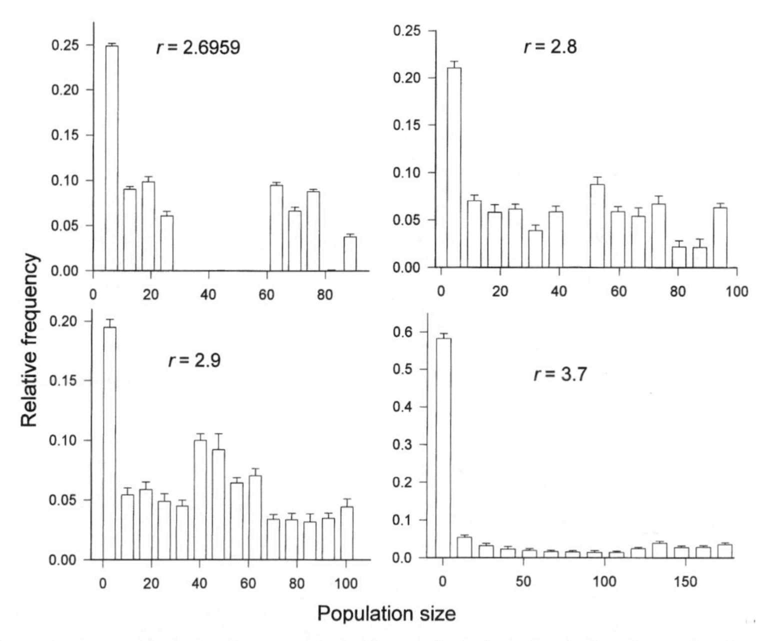

Fig. 2.2 Average relative frequencies (with standard deviations) for 500 replications of time series started with a random number of individuals using the discrete logistic equation for four different values of r. Each replication consisted of 1,000 iterations; $K = 50$ for all simulations.

There is an important consequence of ergodicity for those chaotic systems that have this property: the statistical aspects of stability generated by the system's dynamics are only apparent at relatively large scales. In the discrete logistic example, only after the accumulation of many, relatively long time series was the stability of the statistical distributions evident. This statistical stability can often be identified geometrically (fig. 2.3). The important point here has to do with scale: the regular statistical properties of ergodic chaotic attractors can be discovered only by observing the system at a sufficiently large spatial or temporal scale.

The true message of chaos for biology is this: if we hope to discover regularities or generalities in ecological systems, we must seek them at appropriately large spatial and temporal scales. If we do not, then we will end up looking at the unpredictable, nonrepeating dynamics of individual systems that fail to capture the true nature of the processes we study.

The conclusions drawn in the previous paragraph assume, of course, that populations, communities, and ecosystems are all governed by ergodic chaotic systems. This is an empirical issue, and one that has not been clearly settled by analyses in the past. I will defer the discussion of

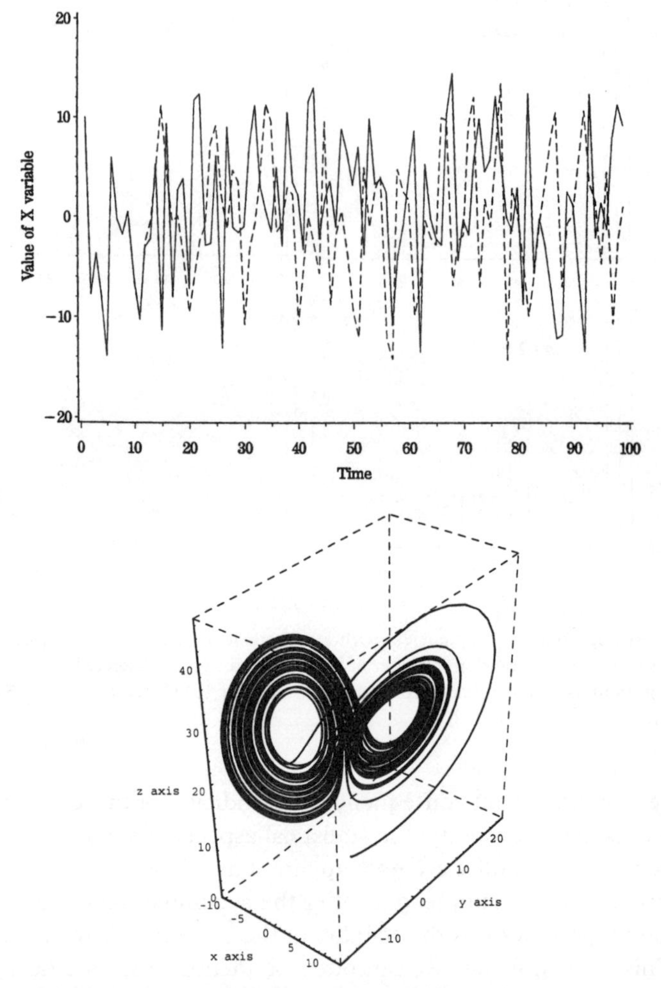

Fig. 2.3 Solutions to the Lorenz system of three differential equations. The equations were initially used to model turbulent flow of air masses in the atmosphere. Time series for two solutions of these equations are given in the top panel. Each series was started using a slightly different initial value for *x;* all other starting values and parameters were equal. Note that the time series diverge. In the bottom panel, the trajectory for one of these solutions is plotted in a three-dimensional phase space. Note that this trajectory appears to be very regular. In fact, for a given combination of parameter values, all solutions, whatever their starting values, must eventually lead to the geometric figure outlined.

this point to chapter 5, where I consider the empirical evidence further. Suffice it to say here that the potential existence of chaos in ecological systems does not invalidate the expectation that ecological systems should show structures or regularities that are the result of relatively general processes, or at least classes of processes about which we might be able to discuss some general principles. The challenge for ecology is to find the appropriate scale, or scales, at which these regularities are evident.

How Statistical Thinking Is Used in Modern Ecology

Statistical Practice in Ecology and Evolution

Biological systems, then, have a special type of statistical causality that is both a cause and a consequence of their hierarchical structure. Interestingly, it is not generally appreciated that many modern theories of biological processes are implicitly structured as hierarchical models with statistical causality. This is partly due to the fact that statistical thinking in biology has mainly been limited to the analysis of data. I therefore review critically the role that statistical practice plays in modern ecology and evolutionary biology, and I examine the limitations of statistical data analysis in studying complex biological systems. I then look at some examples of theoretical constructions in ecology and evolution that are implicitly hierarchical and statistical in their formulations.

With the rise of modern theories of data analysis, an appreciation of the relationship between the variability of biological systems and their orderliness has been replaced with a more reductionist tendency to ignore variability in biological systems and instead focus on properties of the systems which are thought to be ontologically persistent. In such an approach, variability is thought to mask the "true nature" of the system. This approach has been successful for analyzing many systems, but will be effective only when the effects on the system are (1) consistent among different parts of the system and (2) large relative to differences among parts.

This modern approach to biological systems is perhaps epitomized by the use of general linear models. The general form of all such models is

$$\mathbf{Y} = \mathbf{XB} + \mathbf{E},$$

where **Y** is a matrix of "dependent" variables, **X** is a matrix of "independent" variables, **B** is a matrix of unknown parameters that describe the relationship between the dependent and independent variables, and **E** is a matrix of unknown "random" effects.[13] Depending on the type of data in **X** and **Y**, this general model gives rise to statistical procedures as divergent as log-linear models, discriminant analysis, regression, and

13. Some types of data that are inherently nonlinear can be transformed to approximate linearity. Thus, the linear statistical model is more broadly applicable than is implied by its relatively simple form.

analysis of variance (Dobson 1983). Given the data in **X** and **Y** it is possible, using various statistical criteria such as least squares or maximum likelihood, to estimate the unknown parameters and random effects in the general linear model. The most common solution is given by least squares as

$$\mathbf{B}^* = (\mathbf{X}'\mathbf{X})^{-1}\mathbf{X}'\mathbf{Y},$$

$$\mathbf{E}^* = \mathbf{Y} - \mathbf{X}\mathbf{B}^*,$$

where the asterisk indicates an estimate based on the data.

Although the general linear model provides a powerful set of techniques for describing empirical data, there are some major difficulties involved in using these techniques to analyze complex biological systems. These difficulties should not preclude the use of linear models in analysis of ecological and evolutionary data, but should guide their use and interpretation.

Perhaps the greatest difficulty in using linear models in the analysis of ecological and evolutionary data is the confusion they can produce regarding causality. It is often convenient to the argument of the researcher to assume that causal relationships exist among the dependent and independent variables such that the dependent variables "respond" to changes in the independent variables. Levins and Lewontin (1985, 109–22) gave an enlightening discussion of the relationship between analysis of variance estimates of the environmental and genotypic components of phenotypic variation and the real genetic and environmental determinants of that variability. Their point is clear: statistically significant variance components do not imply genetic or environmental causes. Much more detailed information regarding the genetic and environmental systems is required to establish causality. Generally, it is not appropriate to assume that a linear statistical model represents causal relationships among variables. Only when an investigator has great confidence that the actual causal model is linear (or linearizable by transformation) should a linear analysis be attempted.

An example of the problems of inferring causality using linear models was given by Ives (1995). He generated data from linearized (log-transformed) stochastic population models for single and multiple communities and was able to obtain meaningful results for species responses to simulated perturbations using linear regression. As encouraging as his results are, the caveat is that his models were constructed so as to be nearly linearizable by log-transformation. There is no guarantee that linear regression would work with other kinds of models. The larger problem not addressed by studies like Ives's is one of model identification.

Given that we have no knowledge of the actual equations governing population dynamics of species in a community, before we can be certain that a particular solution is appropriate, we need to be certain that we have obtained the best model available to describe the data.

A second difficulty with linear statistical models is related to the first. If a linear model is not a causal model, it can only imprecisely represent the true relationships that exist among variables. Traditionally, it has been assumed that most of this imprecision can be modeled adequately by the "error" terms in the linear model (i.e., the matrix **E**). However, this may not be an accurate assumption in a complex system where nonlinear, autocorrelated effects abound. The probability models used to represent **E** may themselves be only caricatures of "random" effects. These problems give rise to several practical difficulties. First, criteria for statistical independence are probably always violated (see, e.g., Hurlbert 1984). Second, logical difficulties with the interpretations of results, even from controlled experiments, abound (Schaffer 1981; Bradley 1983; Bender, Case, and Gilpin 1984; Yodzis 1988). Third, the relevance of theory to empirical findings becomes tenuous (Pielou 1981; Simberloff 1982, 1983; Wiens 1983). Finally, there is the continual hazard that data and statistical models inappropriate for the scale of the process being studied will be used to develop invalid tests of those processes.

As an illustration of the problems that can be encountered when using linear statistical models to analyze complex ecological and evolutionary hypotheses, consider Schluter's (1986) use of variance components analysis to estimate the role of phylogeny and habitat similarity in determining convergence in community structure of two bird communities on different continents. The hypothesis being tested was that, given similar kinds of species (in this case, seed-eating species from the bird families Emberizidae, Fringillidae, Estrildidae, and Ploceidae), each community should converge on similar "solutions" to the environmental problem of eating seeds in competition with different combinations of species within the ecological constraints of a particular habitat. That is, one should see similar community characteristics in similar habitats after removing effects of phylogeny. Aside from the technical problems of estimating variance components in mixed model analysis of variance with missing cells (Milliken and Johnson 1992), the problem with Schluter's analysis was that he assumed that phylogenetic effects were linearly additive. This is inappropriate because phylogenetic effects are autocorrelative (Harvey and Pagel 1991; Gittleman and Kot 1990; Gittleman and Luh 1992; Garland et al. 1993). Even if estimates of phylogenetic effects are unbiased, any statistical evaluation of them will have inflated type I errors due to underestimated error variances. Logically, phylogeny is causative in that it determines the genetic potential for adaptations to environmental conditions experienced by a species, so even accounting for phylogenetic autocorrelation may not allow testing of the stated hypothesis. Finally, Schluter's data came from censuses of single communities chosen to represent particular habitat types. There was no allowing for the possibility that these communities were not representative of the geographic variation that may have existed in community structure (Wiens et al. 1986). Overall, the problem with this study is that the model (a simple linear statistical model) was inappropriate for addressing patterns determined by a complex set of processes that operate at large spatial and long temporal scales.

What role should statistical data analysis play in the study of ecological and evolutionary systems, given the pitfalls of current statistical practice? One possibility is to view statistical calculations as "filters" that emphasize some patterns in the data at the expense of others. In most data sets, particularly those collected without experimental controls, many factors may contribute to the observed patterns in the data. In performing a statistical calculation, it can be assumed that each calculation places large weights on a small number of causal factors operating on a limited set of spatial and temporal scales, while placing small weights on other factors. The strategy of data analysis is then to apply numerous statistical calculations to the data set in an attempt to find calculations that weight different processes differently (Allen and Starr 1982; Maurer 1985b, 1987). Such a strategy is essentially a problem of model identification. I will come back to this point in chapter 10 and make some suggestions about where to go from here.

Statistical Paradigms in Ecology and Evolution

Current statistical thinking in ecology and evolution is often restricted to a narrow range of problems dealing primarily with data analysis. This has not always been the case. In the following sections, several contributions to biological theory are examined. In each case it is evident that the authors were using statistical descriptions of biological complexity as the basis for their concepts regarding biological structure.

Lotka's Concept of Physical Biology

Lotka (1925) attempted to develop a theoretical structure to elucidate the physical basis of biological phenomena. His was really a theory of statistical causality, and shares much in common with the ideas developed above. Lotka's attempted synthesis of thermodynamics and biology was intended to give physical meaning to biological concepts such as population density in the same way that statistical thinking in physics united the dynamics of individual particles with measurable quantities such as pressure.

Lotka's general analogy was that population density was the statistical outcome of the behavior of individual organisms as they interacted with one another and their environment in the same sense that pressure was the statistical outcome of the collisions of individual particles of a gas among themselves and with the walls of a container. Furthermore, since Lotka realized that biological systems were much more complex than mixtures of gases, he hypothesized that the course of biological evolution was ultimately determined by the flow of energy through populations of organisms (Lotka 1922a,b; see also Maurer 1987). This complexity of energy flows eventually led to increased levels of organization in much the same sense as was discussed above. In Lotka's view, environmental constraints were imposed by the availability of energy, and historical constraints by the success of organisms in securing that energy (Lotka 1922a,b, 1925). Lotka did not have the advantage of a clear understanding of the nature of inheritance. He had read some of the initial work of Fisher on the subject of evolution, but had no opportunity to integrate a mature understanding of evolution by natural selection (what he termed *intraspecies evolution*) into his overall framework.

Statistical Thinking in the Neo-Darwinian Synthesis

Statistical thinking played a crucial role in the reconciliation of Darwinian theory with Mendelian genetics. Fisher's (1958) formulation of the fundamental theorem of natural selection was based on the assumption that reproductive populations of organisms were in general very large and composed of a variety of organisms. In Fisher's theory, environmental constraints consisted of a constantly changing set of abiotic and biotic factors that allowed only some organisms to reproduce. Historical constraints consisted only of the current complement of genetic variation inherited from past generations. The dynamics of phenotypic change due to natural selection was then described as the statistical outcome of the fates of many individual organisms. Fisher was impressed with the similarities of his theory of evolution by natural selection to the second law of thermodynamics, although he was also aware of the differences between the two. The fundamental theorem of natural selection states that the rate of change in fitness due to natural selection is proportional to the additive genetic variation in a population. Since the additive genetic variation must always be nonnegative, natural selection cannot decrease fitness. Fisher pointed out the similarity between this result and the idea from thermodynamics that the rate of change in entropy in a thermodynamic system must always be nonnegative. This comparison has been misinterpreted by some evolutionists who understood Fisher's fundamental theorem to indicate that natural selection must result in some advantage to the population (Maynard Smith 1989, 118). This reasoning confuses cause and effect. The theorem describes an effect of a statistical process given certain constraining conditions: *if* there is heritable variation in the population, and *if* that variation is correlated with differences in fitness among different types, then, on the average, as less-fit individuals are eliminated, only more-fit individuals will be left. It should be noted that the theorem does not preclude a maximum fitness from existing. But the change in average fitness does not require that any advantage to the population or species must accrue due to natural selection (Fisher 1958, 49).

Wright (1931) also developed a statistical approach to evolutionary change. His basic approach was similar to Fisher's, but differed in that he assumed that natural populations were divided into small subpopulations, and therefore the ability of organisms to reproduce panmictically was restricted. Nevertheless, he viewed evolution as an interplay between a large number of alternative phenotypes generated by the existing genetic variation in the population and a changing environment. Wright, however, added population structure to the set of constraints governing the dynamics of evolution. Thus, in his view, changes in allele frequencies in populations could result not only from differences in fitness among individuals in the population, but from restrictions on matings imposed by population structure. It is important to note that, despite their differences in content, the theories of Fisher and Wright regarding the dynamics of evolution shared a similar structure.

Statistical Theories of Abundance

In addition to contributing to evolutionary theory, statistical thinking led to the development of several theories regarding the ecological problem of species diversity. With the development of community ecology and competition theory, these statistical approaches were to a large degree replaced with a more reductionist view which assumed that the determinants of species diversity could be understood by examining local communities to determine patterns of interaction among individual species (MacArthur 1972). As mentioned above, Brown (1981) concluded that such attempts had failed to find satisfying generalizations regarding the generation and maintenance of species diversity.

Recently, Rosenzweig (1995) suggested that the oldest ecological pattern was the species-area relationship. He presented data from an 1859 survey of the vegetation of the British Isles that showed that the logarithm of the number of species of plants in the compilation increased with the logarithm of area sampled. Preston (1962a,b) argued that the relationship could be derived from the statistical distribution of abundances among species within a local community (see also Sugihara 1980). Despite these attempts to derive a statistical mechanical theory for the species-area relationship, the problem has remained somewhat resistant to solution. I consider the species-area problem in more detail in chapter 8.

Fisher, Corbet, and Williams (1943) first suggested that the log-series distribution could describe the distribution of abundances of species in an assemblage. Williams (1964) further developed the empirical basis for the log-series distribution. Kendall (1948) developed a theory of population dynamics that, though probably too restrictive to be useful for most ecological systems, gave rise to the log-series distribution. Later, Brian (1953) suggested that abundance distributions could also be described by a negative binomial distribution. Perhaps the most influential work regarding abundance distributions was that of Preston (1948, 1962a,b) who argued that the lognormal distribution was also appropriate for describing species abundance patterns. May (1975) reviewed the statistical properties of these different distributions, and Pielou (1975) showed that some of these distributions are related. A recent treatment of practical uses of these distributions in analyzing species diversity is given by Magurran (1988).

Kerner (1957, 1959) developed an elaborate theory that was intended to be analogous to statistical mechanics in physics for biological systems. He began by assuming a specific form of Volterra's equations for the dynamics of a system of many species. This form assumed (1) that there were no self-damping terms in the equations and (2) that all interactions among species were predatory. Kerner's calculations resulted in some interesting predictions. First, Kerner's theory postulated that over time species' average squared deviations away from their respective equilibria were constant. Second, Kerner derived a distribution of species' deviations from their equilibrium values that was identical, in the case of large average deviations, to the log-series distribution developed by Fisher, Corbet, and William (1943).

Unfortunately, the assumptions underlying Kerner's (1957, 1959) calculations limit the usefulness of his predictions for ecological systems (Maynard Smith 1974). The limitation of interactions to predation and the assumption of no intrinsic population control (indicated by the absence of self-damping terms in the equations) greatly restrict the applicability of Kerner's results. However, it is interesting to note that Kerner's theory hinged on the statistical patterns of species' squared deviations away from equilibrium values. This is very similar to the definition of the variance of species abundance, which is the squared deviation of a species' abundance away from its expected value. If Kerner's specific results are generally not applicable, his approach suggests that analysis of variation of abundance within and among species may shed light on the structure of species assemblages. Ulanowicz (1986, 1997) and Pahl-Wostl (1995) have taken an approach not unlike Kerner's in that they have concentrated on partitioning the information measure of average uncertainty (known to ecologists as the Shannon-Weiner diversity function). There is a close relationship between this measure and the variance.

Commonalities among Statistical Theories in Biology

The statistical theories discussed above all have a common structure. First, they assume that there is a large number of individual organisms in the assemblage being studied. Preston (1948), for example, could not apply his lognormal distribution technique to small samples. The implication was that the processes generating a lognormal distribution of abundance probably operated on large spatial scales (Preston 1962a,b). That is, most samples taken from relatively small geographic areas include many species that probably would not be able to persist within that region if it were not for the fact that the small area was an arbitrary sample from a much larger area. Preston explicitly mentions that this fact probably makes preserving diversity within isolated national parks an exercise in futility. Second, all of these theories suggest that the constraints inherent in the properties of the components of the system (e.g., organisms) and the constraints imposed from outside the system by the environment interact with the large number of organisms in the system to cause regular patterns of abundance (see also May 1975). These essential elements can also be identified in statistical theories of evolution, such as Fisher's fundamental theorem of natural selection.

The constraints inherent in system components such as organisms are largely historical in nature. For example, the history of changes in the genetically determined properties of organisms constrains the ways in which they process energy, and hence their ability to survive and reproduce. On the other hand, the constraints in the environment of a system are often a mixture of historical constraints (e.g., the ecological properties of competing species) and physically determined properties of the earth and its climate. For example, in natural selection, the properties of other organisms that affect the fitness of individuals in an evolving population are determined by evolutionary history, while the properties of weather patterns that may influence fitness seem to be largely ahistorical (even though they might be nonequilibrial or chaotic in nature). The ahistorical nature of weather patterns is due to the fact that there is no mechanism by which the atmosphere stores information about its history.

Prospects

Not all biological problems are amenable to analysis as large number systems. Problems such as phylogeny reconstruction or the origin of a particular species or clade seem to fall into this category. In addition, ecologists and evolutionary biologists who study the adaptations of individual organisms must consider the nature of each of a small number of individual organisms. Slight differences among organisms may influence the nature of their interactions with other organisms and with their environments, and these differences must be preserved in any meaningful description of such a system. Clearly, a statistical approach to such problems is inappropriate.

Despite the widespread acceptance of such reductionist techniques as are currently used in the study of adaptation, there is the expectation that important phenomena exist at other levels, some perhaps yet undescribed, that may only be revealed by a statistical approach. The earlier success of statistical approaches in the description of abundance distributions was tempered by many examples of how species abundance distributions such as the lognormal or log-series distributions could be generated by sampling models that required no biological mechanisms (e.g., Gotelli and Graves 1996). This is an important criticism of these early approaches to statistical mechanics of ecological systems. The problem, however, is really one of model identification. The fact that a sampling model can replicate a species abundance distribution gives such a model no necessary logical priority. The issue would be much better resolved by data. For a given data set, the model that *best replicates* the empirical pattern would be the preferable model. What we need, then, are objective criteria that allow us to differentiate between competing, complicated models (Hilborn and Mangel 1997; chap. 10 in this volume), not dogmatic assertions that sampling models have some kind of necessary priority over other kinds of models.

The bias against statistical approaches in ecology, however, seems to be waning. There is a growing expectation that many problems can be addressed using a statistical approach. Perhaps the most important of these problems is that of the generation and maintenance of biological diversity. Much of the previous work attempting to understand diversity has approached from a reductionist viewpoint, attempting to explain diversity based on the characteristics of individual organisms and their interactions with each other and their environment (elegantly summarized in MacArthur 1972). Brown (1981) has argued that such approaches are inappropriate because they do not address the problem of diversity at the right scale.

I suggest that to understand diversity at biogeographic scales, it is necessary to adopt a much broader approach. Because there are literally billions of organisms in most floras and faunas, James Brown and I (Brown and Maurer 1989; see also Brown 1995) proposed that the most appropriate way to study them is to describe and analyze them as statistical systems. Thus, for example, it is important to focus on the macroscopic properties of continental biotas in much the same way that a chemist studies the macroscopic properties of gases, while knowing that the gases are complex, dynamic collections of a large number of individual particles. I will attempt to develop statistical models of the dynamics and behavior of individual organisms and populations that have as consequences the macroscopic properties of species assemblages. Such macroscopic properties can be, and have been, measured, and I suggest that continued measurement and comparison of such properties will provide much of the data that will eventually lead to a clearer understanding of how the diversity of a biota develops, changes, and is maintained through time.

The empirical success of the macroecology research program described by Brown (1995) is an example of how ecology can be "simplified" by examining phenomena at large scales. Larger scales at which phenomena have been studied include continental biotas, clades of related species, and collections of many local communities. The principles of statistical causality outlined here suggest that (1) in studying macroscopic ecological systems we should be looking for evidence of constraints imposed on these systems from higher levels;[14] and (2) the properties of macroecosystems that are most interesting to study should be statistical properties that summarize small effects of many different, individual organisms (e.g., means and variances in abundance, geographic range size and shape). The simplification introduced by macroecology is that we do not have to study the fates of billions of individual organisms living in very localized ecosystems in order to understand how biological diversity evolves over long periods of time or across large expanses of space. Instead, the natural scale for such investigations is the scale at

14. For example, a biota occurring on a continent would be constrained by the size, shape, and latitudinal extent of the continent. The continent imposes external, larger-scale constraints on the biota because the biota is largely contained within the continent.

which entire species interact with continental and global ecosystems and with each other over evolutionary time. Macroscopic measures of ecosystems can be of two types. First, there may be structures that can only be defined by the ensemble as a whole. An example might be the trophic structure of a continental ecosystem or the size of the geographic range of a species. Second, structures may be defined by the statistical properties of the ensemble. Such measures might include the average rate of nutrient cycling across a landscape or the average body mass of adults in a species.

Another benefit of the principles described here is that they suggest new opportunities for investigating ecological processes and phenomena across scales. Generally, when we attempt to move from larger, more inclusive scales to smaller ones, we should be looking at how processes at the larger scale constrain those at smaller scales. Conversely, when we attempt to move from smaller to larger scales, we should be attempting to show how processes at the smaller scales determine the statistical properties of entities at larger scales. The mode of explanation is different in each case. In the former, we look for boundaries or types of processes that cannot occur. In the latter, we attempt something akin to classical statistical mechanics: we try to show how complexity at the lower scale leads to regularities at higher scales. I will be emphasizing these themes throughout the rest of this book.

CHAPTER THREE

Communities on Small Spatial and Temporal Scales

> Humpty Dumpty told Alice, "When *I* use a word, it means just what I choose it to mean—neither more nor less." Irrespective of how other people use the word "community"—and there are almost as many uses as there are ecologists—I use it here to mean any set of organisms currently living near each other and about which it is interesting to talk. The question is not whether such communities exist but whether they exhibit interesting patterns about which we can make generalizations.
>
> R. H. MacArthur (1971)

Community ecologists have been very sloppy about what they have called communities in the past. MacArthur's assertion quoted above is perhaps the epitome of this sloppiness, but there is a good reason for it. As I pointed out in chapter 1, early in the development of ecology it was thought that there were discrete, recognizable assemblages of species that were consistently found together and that interacted so closely as to form a kind of highly organized association that shared properties similar to an organism. If one believes recent textbooks, this view has pretty much been laid to rest (see, e.g., Begon, Harper, and Townsend 1990). In the overwhelming majority of cases, it appears that although species do form complex networks of interactions in ecosystems, these networks are highly variable (Yodzis 1988). Therefore, if boundaries among groups of interacting organisms are not very distinct, any definition of a community is likely to be arbitrary. The sloppiness that one might choose to ascribe to community ecologists in defining the object of their study comes not from some inherent laziness or lack of perception on their part; rather, it comes from the inherent haziness of the systems that they study.

Yodzis (1988) gives theoretical justification for this nebulous view of local community "structure." In a study of plausible community matrices constructed from published food webs, Yodzis determined that net interactions of species with one another were indeterminate in both sign and magnitude. That is, slight changes in parameter values resulted in very different sets of effects of species on one another. This indeterminacy showed up not only in the signs of the net effects of species on one another, but on the topology of their net interactions (i.e., which species affect which other species).

Despite the fuzziness in the definition of communities, at least on the scales at which ecologists commonly study them, experiments in natural systems designed to identify ecological mechanisms of interaction have been very successful in demonstrating that processes such as competition do indeed operate among species in the study plots where the experiments take place. Early reviews of the literature provided somewhat equivocal evidence for the importance of species interactions in determining community structure (Schoener 1983; Connell 1983). More recently, rigorous examinations of studies that were sufficiently replicated to provide a measure of statistical reliability have unquestionably provided support for the importance of species interactions (Gurevitch et al. 1992; Menge 1997). Yodzis (1988) points out that the interactions documented in short-term experiments may change in sign and magnitude over time in a complex community. Relatively long-term (i.e., decade-long) experiments have shown that the nature of species interactions does indeed change over time (Brown et al. 1986; Brown, Valone, and Curtin 1997), and that many of these changes can be attributed to changes in environmental conditions such as climate. Paleontological studies of quaternary communities suggest that many communities have undergone significant reorganization in the past several thousand years (Graham 1986; Delcourt and Delcourt 1991). If this tendency of species interactions to change over time is universally true, that implies that the interactions documented in localized, short-term experiments are likely to be transitory states that will continue to change over time, and probably across space as well (see, e.g., Wiens 1986; Brown and Kurzius 1987, 1989).

In this chapter, I consider the limitations of conducting experiments in local communities. There is much to be learned by maintaining a rigorous experimental research program in ecology. Many ecological mechanisms can only be studied in this way. But given the complexity of ecological phenomena and the limitations of conventional statistics in describing them, there is a danger that a localized experimental research program will miss important phenomena that operate in macroecological systems at scales undetectable by any reasonable experiment. The fact is that humans are currently conducting an unconscious experiment on a global scale that is uncontrolled and has many unforeseen consequences (see, e.g., Turner et al. 1990). Because of the connectedness of the earth's ecosystems, the effects of this experiment are difficult to understand within the context of a single, experimental system (Schneider and Boston 1991; Pace 1993; Ehrlich 1997). What is the optimal role for experimental community ecology within the context of human-induced global change? What can this research program contribute, and what are its limitations? The answers to these questions will make clearer the role and limitations of the complementary macroecological approach to ecological problems that I consider in later chapters.

The Importance of Studying Local Communities

What is to be learned from studying local communities within the broader context of a changing, dynamic biosphere? On the one hand, Ehrlich (1997) argues that isolated studies that replicate some experimental protocol to establish the details of the interactions among a group of organisms within a local system have little likelihood of providing any new insights into how ecosystems operate. On the other hand, a reductionist view of causation is cautious of accepting as evidence any information that is not obtained from carefully controlled, appropriately manipulated experiments. From this viewpoint, cause and effect relationships are difficult to establish in any other way. These contrasting views sometimes lead to an impasse between "holistic" and "reductionist" ecologists.

Consider how the holistic and reductionist approaches might examine the same problem. Suppose that two ecologists, a holist and a reductionist, are each asked to submit a proposal to study the impact of global change on a community of butterflies in a national park. The holistic ecologist might propose to study the geographic ranges of the species involved, examining population dynamics within the park relative to where in each species' geographic range each censused community was located. The ideal conditions for each species would be determined by careful study of what factors limited populations at each species' geographic range boundaries. Predictions regarding the response of communities to global change might be made by identifying predicted changes in environmental conditions from global change models and comparing them to the conditions that limit the range boundaries of species. The reductionist might plan a series of experimental manipulations that would identify the specific factors in each study site that determined the abundance of each species. Experimental sites might be constructed along environmental gradients (e.g., a moisture or elevation gradient), and the outcomes examined in light of changes in the gradient itself. From these results the reductionist would infer what community changes might occur as a consequence of changes in local conditions that would be analogous to differences among sites along the gradient.

Which of the two approaches outlined above would provide the best, most reliable information about how butterfly communities within the park would change? Which of the two types of studies might be generalizable to the study of vertebrates in the park? Which would be most appropriate to examine butterfly communities in other parks? Although one might argue that one or the other approach should be used, I think that a third, more profound alternative exists. *Both* approaches contribute information about the communities being studied. In fact, they provide *complementary* information. The experimental studies might be used to design specific management plans for the butterfly communities in the park, while the geographic studies might suggest how to apply the experimental results to other groups of organisms or to the same types of communities in other parks. Rather than competing for funding (and journal space), the two approaches to butterfly communities should both be funded (and published). The information obtained from one kind of study without information from the other gives an incomplete view of the community.

It should be noted that reductionist studies are not necessarily synonymous with "small-scale" studies. Indeed, biogeography often seeks to understand mechanisms as much as does community ecology. Even the most ambitious experimental studies (Carpenter et al. 1995), however, are limited to relatively small spatial and short temporal scales. I would argue that experiments are much more powerful when done within the context of larger-scale comparative studies, and vice versa. What is needed in ecology is coordinated programs that simultaneously examine communities from multiple scales (Wiens et al. 1986, 1993; Wiens 1989b; Kotliar and Wiens 1990; Pimm 1991; Levin 1992; Root and Schneider 1995; Wiens 1995).

The multiscale perspective on communities is slowly replacing the view that communities can be understood only by manipulative experiments. In this and the next two chapters, I examine the limitations of smaller-scale ecological studies. These limitations in no way decrease the contributions that experimental, intensive studies have to make to the future of ecology. What they allow us to see, however, is where the larger-scale perspectives can contribute to a more complete understanding of communities. When we better understand what can and cannot be learned from intensive studies of communities, we can better move toward an understanding of what questions can be asked at what scales and why.

The Nature of Local Communities

Statistical Causality and Structure in Local Communities

In order to see more clearly the limitations of intensive, local-scale studies of communities, it is necessary to consider the context within which those studies are done. It should be obvious to most ecologists that this context is a complex system. Establishing cause and effect relationships in complex systems is not a straightforward task, as we saw in chapter 2. If a system is sufficiently large, much of the complexity occurs at sufficiently small scales so as to contribute little to the overall properties of the system. However, intensive ecological studies are usually not sufficiently large, so the context is very important when attempting to understand the outcome of experiments or observational studies done at that scale.

Up to this point, we have not examined any quantitative details regarding the magnitudes of individual effects and total system size. Just how large does a study have to be to show the aggregate, statistical behavior of a community as opposed to the idiosyncratic, detailed behavior of its parts? As there is little precedent in community ecology for asking such questions, it will be useful to digress for a short period and develop an analogy using a physical system. This analogy will shed some light on the relative magnitudes of effects and numbers of components necessary to obtain statistical effects.

Picture for a moment the following experiment. Suppose we develop a computer-controlled device for mixing gases in a glass flask. The delivery device can determine very accurately how much of each gas is delivered and can record that amount for future analysis. Suppose further that we can vary the amount of energy available to the mixture of gases by heating the flask with a heat bath. The heat bath is also computer controlled, and changes in the heat imparted to the flask can be precisely determined and recorded for further analysis. Now suppose that our goal in developing this device is to measure the behavior of the gas mixture at different scales within the flask. To do this, we insert into the flask 1 million very small and sensitive recording devices which consist of tiny plates (maybe a few hundred square angstroms in area) that record the amount of energy in their immediate vicinity. This energy measurement is determined by the velocities and masses of all of the molecules that strike the plate during a brief instant.

Given such a device, we might ask what kinds of data could be obtained. For each recording plate, we would have a time series of energy measurements that would be determined by the amount of energy imparted to the system by the heat bath and the concentrations of each gas present. We would expect the number and velocities of different molecules that struck each plate to be highly variable, so the time series obtained for each plate might be relatively unpredictable. However, when we accumulated the data from all of the plates, we would expect a time series whose properties would provide fairly accurate information on the mixture of gases and the amount of heat imparted to the mixture by the heat bath. In fact, we could conceivably develop very precise laws regarding the relationship between the properties of the accumulated time series and the properties of the gas mixture and the amount of heat in the system. We might even be able to predict the relative quantities of the gases in the mixture given the temperature and the time series for the accumulated energy measurements. Such predictions, however, would not be possible from the measurements obtained from any one plate.

Why should this be the case? Suppose that only about 10^{18} molecules strike each plate at any given instant. Therefore, the proportion of all the molecules in a 1-liter flask that are being measured at any one instant by any given plate is about $10^{18}/10^{26}$, or 10^{-8}. However, the proportion of the molecules present being measured by all plates simultaneously is considerably larger, about $10^{24}/10^{26}$ or 10^{-2}. Even though the dynamics of molecules hitting any individual plate are governed by very precise deterministic laws, there are so many molecules in the total system that the overall contribution of any small subset of molecules to the properties of the entire system is likely to be quite small. The regular properties of the macrosystem cannot be measured at such a small scale.

This example, contrived as it may be, illustrates the general problem faced by community ecologists dealing with communities at local scales. It is likely that in any one community, the number of individuals sampled represents only a tiny fraction of the total number of individuals in the species. This is because, with the exception of a few extremely rare species, the geographic ranges of species extend far beyond the boundaries of the local community sampled by most studies. For example, in a desert grassland bird community that I studied (Maurer 1985b), I sampled roughly 2 km^2 of habitat. The species with the smallest geographic range that I studied (*Aimophila carpalis*) has a geographic range of roughly 2×10^5 km^2, an area five orders of magnitude greater than the region I was able to study. This phenomenon is not limited to bird communities, but can be seen in communities as diverse as those of coral reef fishes (Anderson et al. 1981) and desert rodents (Stone, Dayan, and Simberloff 1996). In general, the entities that compose the local study plots that are the object of intensive studies do not seem to be very large samples of the possible populations that might be studied. This restriction on the spatial and temporal extent of intensive studies is of major concern to ecologists (Gonzalez and Frost 1994; Carpenter et al. 1995; Baskin 1997).

Although it may not be realistic to expect any statistical regularities in local communities, this does not mean that no important biological information can be obtained from studying them. In fact, I contend that it is important to describe the properties of local communities as precisely as possible, and to elucidate the mechanisms that determine the relative abundances of species in them. However, it is not likely that general principles will be revealed by considering such studies in isolation. The general principles can only come from comparative analyses of many studies done using comparable techniques of data collection and analysis or by expanding the scale of the data collection. For example, Carpenter et al. (1995) reviewed experimental manipulations of "entire" ecosystems on experimental units much larger than those used for typical studies. Results from these larger experiments often identify processes involving interactions among system components that cannot be predicted from smaller-scale experiments.

With the understanding that most studies of local communities are done at too small a scale for any statistical regularities to be likely, it is now possible to examine what kinds of questions can be answered by the continued study of local communities. But first, it is necessary to place such studies in the context of the complex systems in which they are carried out. This requires the introduction of a hierarchical conception of ecosystem organization (O'Neill et al. 1986).

Complexity and Hierarchical Structure in Communities and Ecosystems

The communities that ecologists study on their plots and transects are not isolated entities. As pointed out above, a population measured at such a study area is only a small fraction of the entire species being studied. Often such measurements neglect the spatial structure of the larger population of which the measured population is a part. Spatial structure profoundly affects the population dynamics of a species that is distributed unevenly across a landscape (Johnson et al. 1992; Hanski and Simberloff 1997; Hanski 1997; Gyllenberg, Hanski, and Hastings 1997). Furthermore, the introduction of spatial complexity into models of community dynamics can change the outcome of species interactions from what they would be in spatially simple habitats. If community boundaries are constructed in an arbitrary fashion, the sampling units will be essentially open entities analogous to the tiny collecting plates described above. There is simply no way to guarantee that what happens on the study site has anything but a small effect on the fate of the larger system in which it is embedded. Only in the rare instance of a complete census of a community within a spatially isolated habitat island can we be certain that results obtained from that community will provide us with an adequate description of how the community operates.

One way to view such incomplete systems is to place them in the context in which they undergo their kinetics. That context is explicitly hierarchical for communities. I will not attempt to discuss why communities should have hierarchical structure; I refer the reader to more extended treatments of this point by Conrad (1983), Salthe (1985), and O'Neill et al. (1986). I shall instead address the question of the consequences of this hierarchical structure for the kinds of influences that govern the observable kinetics of communities. Suffice it to say that hierarchical structure is the kind of structure most likely to arise from the statistical causality described in the previous chapter (see Levins 1973).

Communities exist as nested, or scalar (Salthe 1989), hierarchies, where individuals are contained within populations, populations within communities, and communities within ecosystems.[15] There are several factors that can affect the dynamics of populations within communities given this hierarchical structure (fig. 3.1). The number of interacting species is directly determined by the population dynamics of each species in the community. But population dynamics are the consequences of the life histories of individual organisms as they feed, reproduce, and survive. Hence, many physiological and behavioral processes ultimately influence the number of species in a community. Species diversity is also influenced by larger-scale processes that occur in the ecosystem, because each species participates in food webs and nutrient cycles. This means that there are many direct and indirect pathways through which species may influence one another's population dynamics. Direct interactions include processes that incorporate the interactions among individuals of different species, such as interference competition, commensalisms, and symbioses. Indirect interactions among species occur through the mediation of other species, such as two prey species sharing several predators. Predicting the role of indirect effects in communities is not straightforward, even if one has knowledge of the direct interactions among species (Stone and Roberts 1991; Abrams 1993; Billick and Case 1994; Wootton 1994).

15. The discussion here is somewhat simplified. For a discussion of some limitations of a hierarchical view of ecosystems in spatially complex landscapes, see Turner, Gardner, and O'Neill 1995.

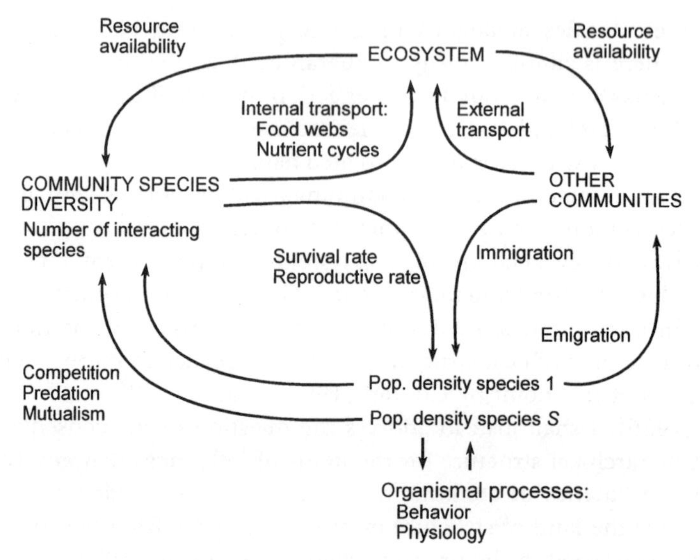

Fig. 3.1 Hierarchical representation of the factors that influence the dynamics of communities. Arrows connect different levels in the hierarchy and show routes of causation. Each arrow represents the net causality of numerous processes, a few of which are listed next to the arrow.

Community diversity is also influenced by larger-scale processes occurring outside the spatial boundaries of the ecosystem. Physical and biological processes cycle nutrients, water, and energy through several ecosystems. For example, the carbon cycle includes both transport processes within the ecosystem and carbon dioxide exchange with the atmosphere. Communities also receive immigrants from populations in other communities in different locations.

The consequence of the existence of complex networks of causality within a community and between the community and the ecosystem in which it is embedded is that the numbers and densities of individual species must vary over time as different influences in the ecosystem affect the population dynamics of each species in different ways. On the one hand, some influences of the ecosystem on population dynamics may be very dramatic, such as an epidemic disease causing the decline of one or more species in a relatively short period of time. On the other hand, there must be many relatively small effects of ecosystem processes on the population dynamics of species, because dramatic changes in relative abundances of species are relatively rare events in most communities. Thus, on empirical grounds, it is easy to justify the expectation that each individual community will be unique. The reason that uniqueness rather than regularity characterizes local communities is twofold. First, even though there may be many species (especially when decomposer trophic levels are considered as part of the community), there are not enough of them to ensure that each contributes only a very small effect on community patterns. Changes in the effects of individual species, especially keystone species, can drastically alter the network of interactions (Yodzis 1988). Second, community boundaries are relatively permeable to effects from outside those boundaries. The dynamics of these metacommunities (Holt 1997) may be significantly different from those we would expect from the community in isolation.

Experimental Community Ecology

Experiments have become a cornerstone of modern community ecology. Experimental demonstrations of competition and predation in laboratory systems formed a major part of population biology from the 1930s until the 1960s and have continued to provide insights into the nature of these interactions in some systems (e.g., Gilpin, Carpenter, and Pomerantz 1986). But the major increase in field experiments since the 1970s provided a wealth of data on species interactions under natural conditions.

Because most communities that have been studied using field experiments are embedded in much larger systems, there are limits to what can be learned from a single, localized experiment. Experiments have clearly demonstrated that competition and predation are important in determining the abundance of species in local communities (Sih et al. 1985; Hairston 1989; Gurevitch et al. 1992). But how far can the results of an experiment on a single community be extended in space and time? This question deserves more attention than it has received, but initial indications are disappointing.

One way of examining the limitations of experimental community ecology is to compare the results of many experiments done on a variety of competitive systems. Fortunately, statistical methods are now available for making such comparisons. Meta-analysis is an emerging field of statistics designed to make statistical comparisons among different studies (Gurevitch and Hedges 1993; Osenberg, Sarnelle, and Cooper 1997). A meta-analysis requires each study included in the comparison to have a mean effect size and estimates of standard errors of the quantities being compared. For competition experiments, this means that the changes in densities of species must be given along with estimates of the standard error of those changes (Gurevitch et al. 1992).

Gurevitch et al. (1992) examined the outcome of a number of competition experiments conducted in the field that had sufficient information to perform a meta-analysis of the results. They analyzed the results of

93 separate experiments varying in duration from a few months to three years. They used a number of strict criteria for choosing which experiments to compare, although there are many sources of extraneous variation among studies that affect such comparisons. The aspect of these studies that I examine here is how the results of these experiments vary over time. After examining the data I will explain why I found the time effect so interesting.

One of the important results that Gurevitch and her colleagues obtained when examining competition experiments was that there seemed to be a decrease in the variability of effect size with the duration of the experiment (fig. 3.2A). That is, only the short experiments showed appreciable effects. Now the effects that they reported were "raw" effects, that is, they were effects calculated directly from the data. Each effect

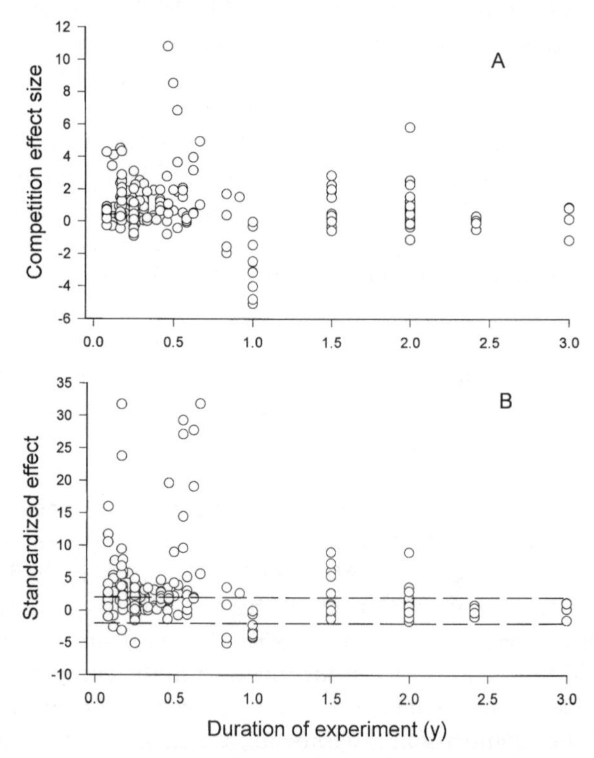

Fig. 3.2 Relationship between duration of the experiment and experimental effect size for experiments conducted on competition reported in a meta-analysis by Gurevitch et al. (1992). *A,* raw effects. *B,* standardized effects, obtained by dividing the raw effect by its standard deviation. Effects above and below the line are greater than two standard deviations away from zero. Note that there are few experiments longer than a year that showed a significant effect, and none longer than two years showed any effects.

calculated from an experiment can be considered a random variable, and an estimate of the variance of these random variables was given by Gurevitch et al. (1992). I took the estimates of these variances that Gurevitch kindly provided and used them to standardize each effect. That is, I divided the effect estimated in each experiment by the square root of the estimated variance (that is, the effect's standard error). These standardized effects are approximately normally distributed, so they can be compared to the standard normal distribution (zero mean, unit variance). Standardized effects less than -1.96 or greater than 1.96 can be considered significantly different from zero.

Standardized effects show a much clearer effect of time. Essentially, no experiment longer than two years showed a standardized effect that was significantly different from zero. In fact, the majority of experiments longer than one year had standardized effects that were not significantly different from zero (fig. 3.2B). One way to interpret this result is that the longer a field study lasted, the less likelihood it had of showing a direct effect of competition. But is this due to some bias introduced by the kinds of species being studied? Looking through the original data set shows no obvious biases. The longest studies involved terrestrial and marine vascular plants, arthropod predators, and freshwater algae. The long-term studies also involved a variety of generation times, ecological circumstances, and types of competition. So no obvious biases can explain the decline in competitive effects observed with experimental duration.

Consideration of the complex nature of the ecosystems in which these experiments were conducted, however, provides a different perspective. Why should the effect of a carefully designed field experiment diminish with time? There are two possibilities. First, the effects of treatments might themselves diminish with time, so that abundances converge to pretreatment levels. Second, variances used to test the effects may increase over time (Pimm and Redfearn 1988; Pimm 1991). The effects may remain the same, but the system becomes so variable that the "noise," or variability of the system, exceeds the size of the experimental effect.

Why should such problems arise in long-term experiments? The answer lies in the fact that field experiments are performed on complex open systems. Although the experimenter can manipulate certain effects, such as the densities of competitors or predators, by definition, the field experiment is conducted in the ecological context of the interactions being studied. The ecological context includes many indirect interactions of the manipulated species with other species and with other processes in the ecosystem. Indirect effects are often resolved on longer time scales than direct effects (Brown et al. 1986; Menge 1997). Furthermore, the experimental treatments are themselves embedded in a changing landscape. Although the experimental effect may dominate the response of experimental populations for a short time, eventually these effects can be swamped by larger-scale effects infused into the experimental system from its ecological context (Brown, Valone, and Curtin 1997). Hence, experimental populations eventually appear to be unresponsive to treatment effects.

An excellent empirical example of such complexities was described by Davidson, Samson, and Inouye (1985). Among other things, they analyzed the response of ants to removal of rodents in a Chihuahuan Desert ecosystem. They could demonstrate no statistically significant effects of rodent removal on ant abundances, although this may have been a consequence of low statistical power due to inadequate replication (two replicates for each treatment combination). However, they did show significant changes in the species composition of annual plants in response to rodent removal. Time series of colony densities for one ant species that were supplied to me by Dinah Davidson are shown for control and rodent removal plots in figure 3.3. Two important points are worth noting. First, for the first two years of the experiment, densities of ant colonies were about twice as high on rodent removal plots as on control plots, suggesting a strong initial increase in ants. However, by the third year, ant

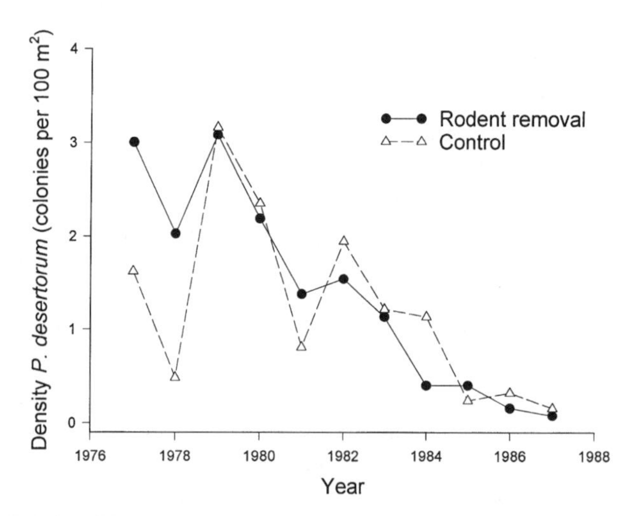

Fig. 3.3 Density of *Pogonomyrex desertorum* colonies in control and rodent removal plots on the Cave Creek bajada near Portal, Arizona. *P. desertorum* is a seed-eating ant and presumably competes with seed-eating rodents. Note that initially there was a large difference between control and experimental plots, but that the difference declined. Note also the drop in density of ant colonies over time.

colonies were nearly equally dense on both control and experimental plots. Second, from the fourth year on, ant colony densities declined, nearing extinction by the eleventh year of censuses. Whatever caused this decline must have operated at a scale larger than the mechanisms that the experiment was designed to measure. This implies a nonclosure of the experiment. The ecological context in which the experimental system resided dominated the long-term responses of both manipulated and unmanipulated populations.

By measuring the responses of plants as well as granivores, Davidson, Samson, and Inouye (1985) were able to address why the response of the ant population to rodent removal was not a straightforward increase (see also Brown et al. 1986). Rodents eat larger seeds than ants. These large-seeded plants compete with the smaller-seeded plants preferred by the ants. Any initial advantages to ants that occurred in response to rodent removal were lost when the large-seeded annual plants began to out-compete smaller-seeded annuals. Rodents indirectly benefitted ant populations by checking the increase of competitively superior large-seeded plants, resulting in more resources available to ants. It is important to note that these interactions were played out in a constantly changing ecosystem. Even though species interactions were important in the short-term regulation of ant populations, larger-scale factors eventually dominated their population dynamics.

After reviewing further experiences with desert granivore communities, Brown et al. (1986) discussed the advantages and disadvantages of field experiments. I concur with the many advantages of experimental community ecology pointed out by Brown et al. (1986) and others (e.g., Hairston 1989). Unfortunately, detailed experimental studies of communities have not led to satisfactory general theories of community structure (Drake 1990; Menge and Olson 1990). General theories of community structure generated before the increase in community experiments during the last few decades have not fared well under experimental examination (Hairston 1989). The experimental approach, with its reductionist, mechanistic perspective, has not yet provided a firm basis for generalization. In fact, as an anonymous reviewer of this chapter pointed out to me, most experimentalists are not interested in generalizations. They are satisfied to work out the details of the mechanisms of interaction within the specific communities that they study.

There are two possible reasons for this lack of generalizability of the experimental research program in community ecology. First, there may be no generalizations possible. That is, the ecological world may simply be too complex to allow for any simplification. In this case, community ecologists are merely clerks cataloging all the unique phenomena that occur in communities over space and time. The second possibility is that generalities might not exist *at the scale of the local study.* Because field experiments are necessarily conducted on open systems, and these systems have relatively few components in comparison to the number of possible components, the systems studied by experiments may be inherently structureless. If structure arises in complex systems via statistical causality, the entities studied in field experiments might be structureless because they do not have the properties necessary to develop structure.

Despite the difficulties inherent in deriving generalities upon which an empirically rigorous theory of community ecology might be built, experimental work in community ecology will continue to be a rich source of information about the details of mechanisms of interaction among different species. Herein lies the importance of the approach. The rich diversity of modes of interaction, both direct and indirect, linear and nonlinear, will continue to provide the basic data upon which our ideas about communities must be founded. Theories will come and go, generalizations proposed and refuted, but the data collected by experimentalists will remain, and any generalization in ecology that is to be useful must be consistent with these data.

The question that naturally arises from considering the results of community experiments is whether or not there even *should* be any attempts to generalize from the results of local-scale experiments. If many experimentalists are content to continue conducting experiments without being concerned with generalizations, the question is whether empiricism alone is sufficient to provide all the knowledge we need in order to react to and mitigate the many changes that humans are imposing on the earth's ecosystems. I think that this is a question that needs to be examined before I proceed.

Conceptual Agnosticism in Community Ecology

One of the results of the increasingly experimental emphasis of community ecology is a kind of "conceptual agnosticism," in which many researchers have begun to look at any theoretical investigations of biological communities with increasing skepticism. For example, in reference to Hutchinson's (1959) suggestion that there are regular size ratios between competitors in communities, Hairston stated, "It was some time before weaknesses in the original hypothesis and in the theoretical edifice built on it were appreciated. Eventually, the importance of predation became apparent . . . and an increasingly skeptical attitude . . . has led to a situation where a rethinking of much ecological theory is necessary" (1989, 16). This skepticism has in many instances had the positive effect of forcing experimentalists to consider more carefully a wide variety of hypotheses, including so called "null" hypotheses (Gotelli and Graves 1996). Because of this, the conclusions made about the outcome of individual experiments have been strengthened. This, in turn, has improved the quality of the database for community ecology. But it has also led to an attitude that theoretical work in ecology is less relevant to our understanding of ecological problems than the empirical research program.

Peters's wide-ranging *Critique for Ecology* (1991) identifies many problems with current theoretical and empirical ecology. In some ways, this book epitomizes the conceptual agnosticism that has permeated the field. Peters argues that past theory in ecology has attempted to explain ecological phenomena using reductionist and mechanistic concepts. Predictions from these theories can be confirmed by observation, but failure to predict adequately the response of a system is not grounds for rejecting such theories. Most of these theories were intended to be general, and in being so lose precision (Levins 1966). Precision is necessary for prediction, however. So, if a general theory cannot adequately predict phenomena, how are such theories to be evaluated? Some have suggested that, to be valid, a general theory need only provide a mechanistic understanding of causal relationships between entities. The problem that arises from this view is that two general theories might provide very different causal explanations for the same phenomena. Thus, explanation is not sufficient to find to "best" theory. One criterion that might be used is Occam's razor, or parsimony, which states that the simplest explanation is the best. There is no reason, however, that simple explanations should be better than complex ones, especially if we know that the phenomena being studied are, in fact, complex. This leads to the paradox that the only way to differentiate between general explanatory theories is by their predictions; but such theories, being general, cannot produce precise predictions.

Sometimes this paradox does not arise in practice because qualitative differences between the logical outcomes of two theories are different enough to allow the use of simple qualitative observations to falsify one theory or the other. But if this were always true, then ecology would not be in the "crisis" that Peters (1991) claimed it is in. Although one might argue with some of the particulars of the evidence that Peters used to justify this claim, I think that most ecologists would see at least some problems with the way ecological science is currently conducted. For example, in a recent exchange on the Internet sponsored by the National Center for Ecological Analysis and Synthesis (NCEAS), James Brown (1997) argued that there have been relatively few conceptual advances in ecology in the past twenty years, compared to the proliferation of important ideas in the 1960s and 1970s. He attributes this to the limited ability of experimental science to understand complex systems by "dissecting" them into their parts. Peter Kareiva's (1997) response to this claim was to argue that ecology is more mature because its analytical toolbox is more sophisticated. In Kareiva's eyes, the future of ecology lies not in finding generalizations or universal laws, but in applying the analytical advances of the past twenty years to real ecological problems. The crisis that Peters talked about seems not to have diminished significantly since the publication of his book.

Peters's (1991) answer to the dilemma in ecology is to cease to construct explanatory, mechanistic theories and focus on the construction of "predictive theories." This he calls "instrumentalism," by which he means that theories are to be viewed as tools that can be used to predict and manipulate the environment. Theories need not explain anything, under this view, but need only predict. Although I agree that prediction is an important component of a successful theory, I am certain that many ecologists are uncomfortable eschewing any attempts to understand why predictions might work. The danger with instrumentalism is that it is possible to construct valid predictive theories based on correlational evidence alone. In chapter 2 I argued that such correlational models are a far cry from establishing causal relationships. Peters is overly idealistic if he assumes that a predictive theory will never be used to infer causality. The temptation is just too great, especially if one has a very accurate model. Pickett, Kolasa, and Jones (1994), after a thorough description of what a scientific theory is, pointed out that such "empirical theories" as advocated by Peters are not theories at all because, although they use assumptions borrowed from established ecological thinking, they avoid "dealing with the conceptual issues that underlie the use of such assumptions" (78). Such approaches cannot lead to an integrated or coherent body of knowledge because they lack the conceptual rigor of true theories.

A very important point that Peters (1991) briefly mentioned, and that I wish to emphasize here, is that predictions and causal connections are scale dependent. Perhaps some of the problems that have plagued ecological theory in the past might become clearer if ecologists were more sophisticated in their development of theories that explicitly included scale. This is the basic message that Allen and Hoekstra (1992) developed in greater detail. I have argued in this chapter that ecological systems at the local scales commonly studied by community ecologists appear complex because the entities being studied are parts of much larger systems. When we study local communities, we see the idiosyncrasies of individual situations. We fail to see the constraints that operate on the larger systems of which local communities are parts. In later chapters, I will consider some of the larger systems that contain communities, and how we might approach them from a theoretical, mechanistic perspective that explicitly incorporates considerations of scale.

It may be that coherent ecological theories may yet arise from the study of local communities. Certainly, the reformulation of fundamental ecological concepts from a more sophisticated viewpoint, such as Leibold's (1995) recent reevaluation of the niche concept, will contribute to the development of better theories for local communities. But I believe a more mature approach to understanding ecological phenomena from different scales (called for in Ricklefs 1987; Wiens 1989b; Levin 1992; Brown 1995) will be more successful in helping ecologists make conceptual and methodological advances than will exclusively relying on detailed, intense, experimental manipulations of local ecosystems.

Conclusions

The ecology of local communities represents a rich source of data and ideas that ecologists should continue to mine in the future. There are conceptual and empirical advances that can be made by studying them further. Problems that have arisen in the past in studies of local communities should not prevent ecologists from continuing the rigorous study of them using theoretical and experimental tools. Nevertheless, there will be limitations imposed on our view of the ecological world if community ecology remains focused only on the local scale. My intent in later chapters is to deal with some of the potential insights that can be gained from expanding the scale of community ecology to connect it with other disciplines. Before doing so, however, I look further into the opportunities and limitations of focusing on local communities in the next two chapters.

CHAPTER FOUR

Communities as Linear Systems

> It has been one of the most agreeable tasks of modern science to trace the wonderfully exact adaptations of the organization of animals to the physical circumstances amidst which they are destined to live. From the mandibles of insects to the hand of man, all is seen to be in the most harmonius relation to the things of the outward world, thus clearly proving that *design* presided in the creation of the whole—design again implying a designer, another word for a Creator. It would be tiresome to present in this place even a selection of the proofs which have been adduced on this point. The Natural Theology of Paley, and the Bridgewater Treatises, place the subject in so clear a light, that the general postulate may be taken for granted. The physical constitution of animals is, then, to be regarded as the nicest congruity and adaptation to the external world.
>
> [R. Chambers] (1844)

The above quote, published anonymously by Robert Chambers prior to Darwin's *The Origin of Species,* recounts the prevailing view of nature in Darwin's day: that of a well-designed machine that operates according to precise laws. To Chambers and naturalists of his day, the exquisite adaptations of organisms and how they worked together were considered things of beauty. Conservationists and environmentalists still commonly portray the natural world as a fine-tuned machine. Such thinking is reflected in phrases like "the balance and harmony of nature." Although much more sophisticated in its scope, the community ecology developed by Hutchinson, MacArthur, and their colleagues was based on the assumption that the population densities of species within communities did not change significantly over time, that is, that they were balanced near a stable equilibrium. Density fluctuations about equilibria were thought to be a relatively unimportant part of community structure, although some ecologists realized that fluctuations in abundance could explain at least some anomalous observations (Hutchinson 1961). The mathematical formulation of twentieth-century community ecology gave the impression of balance and precision.

The equilibrium assumption was derived from the results of Lotka, Volterra, Gause, and others in the early decades of the twentieth century (Kingsland 1995). The assumption of equilibrium allowed community ecologists to treat communities as linear systems. This had the tremendous practical value of allowing the development of a tractable mathematical approach to community description, and made it possible to make quantitative predictions about community structure.

In this chapter, I devote attention to the empirical problems of examining communities as linear systems. It is not entirely clear how we should interpret the linear description of communities or obtain estimates of their parameters. Many methods have been devised for obtaining estimates of these parameters (e.g., Emlen 1978, 1984; Hallet and Pimm 1979). Here I consider a different approach built on the empirical patterns of covariation among species, and I examine the utility of the linear description of communities. To do so in a rigorous fashion would require the examination of a large number of data sets using the same statistical methodology. There is a paucity of data sets appropriate for such an endeavor, so instead I outline the technical issues in generating linear descriptions of local communities and illustrate the use of the technique with a long-term data set on rodent populations. The conclusions I draw from this exercise, of course, are limited to that particular data set. However, the example that I discuss shows both the strengths and limitations of the approach. Ultimately, long-term data on other communities will be needed to determine whether the conclusions presented here are generalizable.

Lotka's Community Model

Before I discuss methods of analyzing community dynamics, it will be useful to look in a little more detail at how Lotka orginally framed his model of interspecies evolution. This model formed the basis of much of MacArthur's argument regarding communities. The context in which Lotka placed his model is often misunderstood, so a brief discussion of his ideas is warranted here. An appropriate understanding of how he viewed his model is particularly important in attempting to understand the empirical meaning of the model's parameters.

Lotka (1925, 50) divided up the study of biological systems into two general fields of study (see Maurer 1987). The first he called stoichiometry, after the use of the term in physical chemistry. This field, according to Lotka, concerned both what he termed kinetics, the mathematical description of changes in biomass (or density) of the components of the biological system, and what he termed statics, the study of the steady states or equilibria described by the kinetics. The second field of study Lotka recognized was what he called the energetics of evolution. This field concerned itself with studying the mechanisms that caused the changes in biomass described by the system's kinetics. Therefore, Lotka separated descriptions of the dynamics of a biological system from descriptions of the mechanisms causing those changes. The parameters in Lotka's equations were not meant to describe mechanisms of interactions among system components, but were meant only to provide a description of the changes in those components. This distinction has been overlooked in many uses of his equations. For example, considerations of "higher order" or "indirect" interactions assume that the parameters in Lotka's differential equations represent causal effects (Stone and Roberts 1991; Billick and Case 1994; Wootton 1994).

Lotka's equations can be used to describe the changes in population density of interacting species as follows. Let the density of the ith species in the community be N_i. Assume that the per capita rate of population change of the ith species is a function of the densities of all species in the community such that

$$\frac{dN_i}{dt} = N_i f_i(N_1, N_2, \ldots, N_s), \tag{4.1}$$

where s is the number of species in the community. Lotka (1920) assumed that there existed equilibrium values for the densities of all species, $N_1^*, N_2^*, \ldots, N_s^*$, that when substituted into the per capita rate of change functions gave

$$f_i = 0, \qquad i = 1, 2, \ldots, s.$$

If such values exist, then the unknown functions f_i can be approximated by a Taylor series expansion about equilibrium densities. Thus

$$\begin{aligned} f_i(N_1, N_2, \ldots, N_s) = {} & f_i(N_1^*, N_2^*, \ldots, N_s^*) \\ & + \sum_{j=1}^{s} \delta_{ij}(N_j - N_j^*) \\ & + O(N_1 - N_1^*, N_2 - N_2^*, \ldots, N_s - N_s^*), \end{aligned} \tag{4.2}$$

where $\delta_{ij} = \partial f_i / \partial N_j$, evaluated at equilibrium densities. The term $O(N_1 - N_1^*, N_2 - N_2^*, \ldots, N_s - N_s^*)$ represents higher-order terms in the Taylor series expansion. Since these higher-order terms are expressed as powers of deviations away from equilibrium densities, the assumption that the community is close to equilibrium ensures that the higher-order terms are close enough to zero to disregard, at least as a first approximation. The constants δ_{ij} are sometimes called competition coefficients (Pielou 1977) or, more generally, interaction coefficients, to include other possible types of biotic interaction. This term implies that these coefficients say something about the mechanisms that underlie the changes of species' population density. A strict interpretation of Lotka's original model (1920, 1925) suggests that this is inappropriate. Two different mechanisms of interaction may result in the same kinetic description of population changes. Likewise, two similar mechanisms of interaction could conceivably result in very different kinetics. Hence, the coefficients should not be expected to provide exact information about the mechanisms by which species are interacting.

Ignoring higher-order terms in the Taylor series expansion and substituting N_i^* for N_i, an approximation to equation (4.1) is then

$$\frac{dN_i}{dt} = \sum_{j=1}^{s} \delta_{ij} N_i^* (N_j - N_j^*), \tag{4.3}$$

or, since $dN_i^*/dt = 0$,

$$\frac{d(N_i - N_i^*)}{dt} = \sum_{j=1}^{s} \delta_{ij} N_i^* (N_j - N_j^*). \tag{4.4}$$

Equation (4.4) represents a system of s linear differential equations that can be written in matrix form as

$$\mathbf{d}(\mathbf{N} - \mathbf{N}^*) = \mathbf{diag}(\mathbf{N}^*)\mathbf{D}(\mathbf{N} - \mathbf{N}^*),$$

where $\mathbf{d}(\mathbf{N} - \mathbf{N}^*)$ is a column vector of rates of change of species' deviations away from their equilibrium densities, $\mathbf{diag}(\mathbf{N}^*)$ is an $s \times s$ diagonal matrix with equilibrium densities on the diagonal, $\mathbf{D}$ is an $s \times s$ matrix containing the interaction coefficients δ_{ij}, and $(\mathbf{N} - \mathbf{N}^*)$ is a column vector of species' deviations away from their equilibrium densities. Substituting $\mathbf{A} = \mathbf{diag}(\mathbf{N}^*)\,\mathbf{D}$ gives the linear system of differential equations

$$\mathbf{d}(\mathbf{N} - \mathbf{N}^*) = \mathbf{A}(\mathbf{N} - \mathbf{N}^*). \tag{4.5}$$

The matrix $\mathbf{A}$ is referred to as the community matrix (e.g., Levins 1968; May 1973).

The elements of the matrix $\mathbf{A}$ are generally interpreted as representing interaction coefficients. As pointed out above, this is inappropriate. The elements of the matrix $\mathbf{A}$ represent the derivative with respect to one species of a general function describing how the abundance of one species changes with the abundances of other species. Whatever processes might be causing those changes are not explicitly specified by the elements of the matrix. Processes that determine the signs and magnitudes of the elements of the community matrix include not only those processes that determine how species actually interact, but also any processes that affect the size of equilibria. For these reasons, interpreting the elements of the community matrix as indicators of process is a misapplication of the model. Discussions of mechanisms require information about specific systems, such as that obtained from experiments. Note that there is another complication in translating results from experimental data to model parameters. Whatever effects species have on one another cannot be interpreted without considering the effects of all other species in the community as well (Billick and Case 1994; Wootton 1994). The exception to this rule is the case where interactions among species are all linear.[16] In this case, the partial derivative of f_i with respect to species j will only contain terms that include species j. If there are significant nonlinearities then terms will be included that reflect the effects of other species on the interaction between species i and j.

It is important to point out the relationship between the equations defined in equation (4.4) and the more common formulation of the Lotka-Volterra equations in general use. First, substitute N_i for N_i^*. Then define the quantities

$$\alpha_{ij} = \frac{\delta_{ij}}{\delta_{ii}},$$

$$r_i = -\delta_{ii} \sum_{j=1}^{s} \alpha_{ij} N_j^*,$$

$$K_i = \sum_{j=1}^{s} \alpha_{ij} N_j^*$$

$$= N_i^* + \sum_{j \neq i}^{s} \alpha_{ij} N_j^*.$$

Substituting these into equation (4.4) gives

$$\frac{dN_i}{dt} = r_i N_i \left(1 - \frac{N_i + \sum_{j \neq i}^{s} \alpha_{ij} N_j}{K_i} \right). \tag{4.6}$$

Note that $\alpha_{ij} = 1$. This is the textbook formulation of the competition equations. The meanings of the constants in this formulation are made

16. Linear means that f_i can be written as a sum of constants multiplied by the densities of each species, that is, $f_i = \sum a_{ij} N_j$.

clearer by considering the substitutions. First, the "competition" coefficients, α_{ij} are the relative effect of species j on species i compared to the effect of species i on itself. Here, "effect" is defined precisely as the partial derivative of the function f_i, the per capita rate of growth of species i, with respect to species j, evaluated at equilibrium densities. The term "effect" does not imply a specific mechanism of interaction and, as just mentioned, includes the effects of all other species on species i as well if there are nonlinear terms in f_i. Second, r_i is the negative of the effect of species i on itself times the sum of the relative effects of all other species on species i weighted by their equilibrium densities. The variable r_i is typically referred to as the intrinsic rate of increase of the species. If only one species was in the community, then this interpretation would be satisfactory. But the fact that interaction coefficients and equilibrium densities of other species are explicitly included in the definition of r_i implies that the rate of growth of any population that is interacting with other species is never independent of other species, but is affected by a common state (equilibrium) determined by the equilibria and interaction network among species. Finally, the constant K_i, often called the carrying capacity, is the sum of equilibrium densities of species in the community weighted by their relative effects on species i. The behavior of a Lotka-Volterra competition system is determined by the nature of the equilibrium densities of the species and the way that densities of species affect the changes in densities of other species.[17]

If the assumption of linear effects is justifiable, then the solution to equation (4.5) depends on the eigenvalues of the matrix **A.** For the ith species, its density as a function of time is

$$N_i(t) = N_i^* + \sum_{j=1}^{s} c_{ij} \exp(\lambda_j t), \tag{4.7}$$

where λ_j is the jth eigenvalue of the matrix **A** and c_{ij} is a constant obtained from the eigenvectors of **A** (Lotka 1925; Puccia and Levins 1985; Edelstein-Keshet 1988; Yodzis 1989). The behavior of the solution is determined by the nature and signs of the eigenvalues. If any eigenvalues are imaginary, then the solution will be periodic. The solution will converge on equilibrium densities only if the real parts of all eigenvalues are negative. If the real parts of eigenvalues are zero, then the solution exhibits sustained, bounded oscillations. If the eigenvalues of **A** describe a solution that converges on equilibrium densities, then they also define a

17. Note that the Lotka-Volterra predation equations cannot be derived from this formulation. This is because predator and prey experience no density dependence, hence the terms α_{ij} are undefined.

locally stable equilibrium to the more general system of equations given by equation (4.1) (Yodzis 1989, 115).

If the functions f_i contain nonlinear terms, then equation (4.7) may be an inappropriate description of community dynamics, particularly when the community is far from equilibrium. However, the eigenvalues of the matrix **A** may still contain some information about community dynamics. The real parts of the eigenvalues are called the Lyapunov exponents (see, e.g., Parker and Chua 1989, 67). The Lyapunov exponents describe the macroscopic structure of the attractor that governs the dynamics of the community. If all are negative, then a simple attractor governs the behavior of the system, and equation (4.7) might provide an adequate description of community dynamics. More often, there will be both positive and negative Lyapunov exponents. This means that the attractor governing the community will have a complicated structure that can generate exceedingly complex dynamics (Schaffer 1985; Gleick 1987). As I pointed out in chapter 2, although such dynamics might appear to be random from the perspective of the dynamics of individual species, the structure of the attractor governing them can be stable if the system is ergodic. Communities governed by ergodic attractors will show macroscopic stability (i.e., persistence of bounded oscillations), while appearing to behave randomly at the microscopic level (i.e., the level of each individual population). It is important to note that the Lyapunov exponents are determined by the nature of the original functions governing community dynamics. That is, the equations describing the dynamics of the community contain information about the macroscopic structure of the community. The complexity of that structure is related to the dimensionality of the attractor described by the Lyapunov exponents. The dimension of the attractor indicates the minimum number of equations needed to describe the community, and hence is a measure of complexity.

The community matrix model of community structure is more complex than the relatively simple formulation would at first appearance indicate. As I pointed out above, the model is often inappropriately applied to biological data. Most importantly, the elements of the community matrix are often misinterpreted as indicating actual mechanisms of interactions among species. A better interpretation of the elements of **A** is to think of them as measures of the *functional dependency* of species on one another. By functional dependency, I mean that the description of the dynamics of any given species depends on a specific mathematical function (f_i). In the case of linear coefficients, that functional dependency is a simple pairwise dependency. If there are nonlinear terms in f_i, then the functional dependency does not express simple pairwise relationships among species.

Under the assumption of linear relationships among species (i.e., the f_i contain only linear terms so the equilibrium densities of species other than i and j do not enter into the partial derivatives in the matrix **D**), the community model still has some complicated dynamics inherent in its formulation. These complications are expressed as the indirect pathways of functional dependencies in the solution to equation (4.5) (Levine 1976; Lawlor 1979). When these indirect pathways are considered for any pair of species, the net interaction obtained by considering all interactions of the two species with other species may actually be of different sign than that indicated by the interaction coefficients of the species (see, e.g., Lawlor 1979). This realization has led to a number of empirical and theoretical attempts to study the importance of these indirect effects.

Lotka's equations, then, provide a powerful descriptive tool for the empirical analysis of communities. The solution to the model includes the effects of functional dependencies of species upon one another. The mechanisms of the dependencies depend on the specific form of the functions f_i in equation (4.1). The magnitudes of the dependencies near equilibrium are expressed by the values that the elements of **A** take on. The difficulty in using Lotka's equations to analyze communities is that the formulation given above explicitly requires that we know the equilibria of each species.[18] In the case of some nonlinear models (e.g., those that are governed by chaotic attractors), it is unlikely that there are any observed population densities that would fulfill the definition of an equilibrium value, even though such values exist for chaotic attractors. Thus, it is necessary to search for a new method of analyzing Lotka's model. In the next section, I consider the relationship between mean and equilibrium densities before introducing a method of calculating the functional dependencies among species.

Mean Densities and Equilibria in Finite Time Series

Biological populations do not exist forever. Most biological populations persist for many years, even centuries, in the same place. But in persisting, most populations fluctuate considerably over time. To what de-

18. One could try to estimate the parameters of the community matrix and use these to estimate equilibria, but recall that the community matrix incorporates *both* functional dependencies and equilibrium values into its elements. It is more straightforward to use the elements of the community matrix to predict the dynamical properties of the community, as is done below.

gree can these fluctuations be accounted for, predicted, or modeled using Lotka's linear system of equations? I will address this question in a moment, but first, I consider an important empirical issue that needs to be addressed before we examine the ability of the Lotka equations to describe ecological populations.

The empirical issue we need to consider here is how we specify equilibria in populations. Since the functions f_i that govern population dynamics are unknown, the equilibria that they specify are also unknown. We are left with the empirical problem of obtaining the best estimate available for these equilibria. We can begin by asking how we might obtain estimates of the values that the f_i take on at particular points in a time series. This is actually fairly straightforward. Appendix 4A shows that the average value that each f_i takes on over a finite time interval, say Δt, is

$$R_i(t) = \frac{\ln N_i(t + \Delta t) - \ln N_i(t)}{\Delta t}, \tag{4.8}$$

where $R_i(t)$ is the average per capita rate of change of the population of species i during Δt. Using values for average per capita rates of change, it is possible to estimate the form of f_i by plotting $R_i(t)$ against $N_i(t)$ (Levins 1979; Puccia and Levins 1985; Royama 1992; Holt 1997).

An example of a plot of average per capita rates of change against estimated population densities for a population of the kangaroo rat *Dipodomys merriami* is given in figure 4.1A (data were provided by J. Brown, University of New Mexico). The first thing to note about the plot is that there is a healthy amount of scatter in the relationship between average per capita rates of change and population density. This scatter is important, and I will discuss its significance in a later section. For the moment, however, assume that the relationship between $R_i(t)$ and $N_i(t)$ can be adequately represented by a straight line with a negative slope (fig. 4.1A). The x-intercept of this straight line is an estimate of the equilibrium population density for *D. merriami,* because it represents the density for which our estimate of f_i is equal to zero. Population density of *D. merriami* clearly oscillates around this equilibrium number (fig. 4.1B).

The surprising thing about the equilibrium value is that, within rounding error, it is identical to the mean abundance of *D. merriami* during the time period. This is true of seven other species in the same community with *D. merriami* during the same time period (table 4.1). These results suggest the following surprising conclusion: in a finite time series, averages provide a very good approximation to equilibria (Puccia and Levins 1985). This conclusion has a hidden assumption. That assumption is that the equilibria themselves do not change over the course of

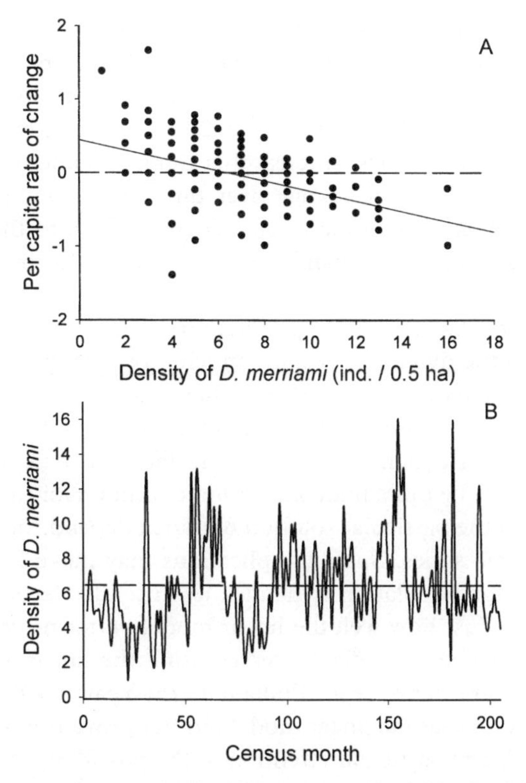

Fig. 4.1 *A,* per capita rate of change of *Dipodomys merriami,* Merriam's kangaroo rat, on control plots on the Cave Creek bajada near Portal, Arizona. Note that although there is much scatter around the line, there is a weak negative relationship between density and the average per capita rate of change. The density for which the regression line predicts a zero rate can be considered an estimate of the "equilibrium" density. Data in this figure represent 17 years of monthly censuses. *B,* density fluctuations of *D. merriami* for 17 years of monthly censuses on control plots at the Portal site. The dashed line represents the estimated equilibrium density obtained in panel A. Note that density fluctuates erratically around this equilibrium.

Table 4.1 Equilibrium and average densities for eight species of Chihuahuan Desert granivorous rodents censused across 205 months. Densities are numbers per 0.5 ha.

Species	Average Density	Equilibrium Density
Dipodomys merriami	6.47	6.49
D. ordii	0.69	0.70
D. spectabilis	1.02	0.94
Peromyscus eremicus	0.32	0.33
P. maniculatus	0.07	0.07
Perognathus flavus	0.15	0.15
P. pennicilatus	0.30	0.28
Reithrodontomys megalotis	0.47	0.48

the time series. Another way of saying this is that the functions f_i are not themselves functions of time, so that the parameters governing the dynamics of species are constants. This assumption is related to the concept of ergodicity introduced in chapter 2. The ergodicity assumption, when applied to a time series, among other things assumes that the mean of the time series does not change over time. This condition actually is met for the *D. merriami* time series. There is more to the ergodicity assumption than this, and I examine the assumption further in the next chapter.

If the functions that govern the population dynamics of species change over time, then the dynamics become much more complicated. For example, Kot and Schaffer (1984) showed that in simple models of population growth, if parameters governing population dynamics change seasonally, a variety of dynamics can result. In fact, for the equations they studied, there may be more than one periodic solution for the same set of parameters, and the particular solution obtained depends on the starting values of the time series. Such complications may make Lotka's linear approximation for population dynamics less useful. Nevertheless, it is important to see just how well the linear model performs with real data sets. There have been notable successes using the linear approach for relatively short time series (e.g., Pulliam 1975; Yeaton 1974), but Pulliam (1985) found that the linear model did not prove useful for longer-term data. I will pay particular attention to the possibility that the linear model of community dynamics can be used on relatively short time scales, but not on longer ones.

Estimating the Parameters of Lotka's Linear Model

Given the similarity between mean densities and equilibria, it is possible that mean densities could be used in place of equilibria in calculating estimates of the unknown parameters in equation (4.5). In fact, by making this substitution, it is possible to show a relationship between these parameters and temporal variances and covariances of densities for each species in a community (appendix 4B). It turns out that if $\boldsymbol{\Sigma}$ is the variance-covariance matrix of densities, then the matrix $\mathbf{A}$ in equation (4.5) can be estimated by

$$\mathbf{A} = \mathbf{diag}(\overline{\mathbf{N}})\mathbf{Z}\boldsymbol{\Sigma}^{-1}. \qquad \textbf{(4.9)}$$

The elements of the matrix $\mathbf{Z}$ (z_{ij}) are

$$z_{ij} = \frac{\int_0^T R_i(t)\,[N_j(t) - \overline{N}_j]\,dt}{T},$$

where $R_i(t)$ is given by equation (4.8) and T is the length of the time series describing community dynamics. One way to think of this is to consider z_{ij} the covariance between the per capita rate of change of species i and the density of species j. Clearly, there is information in this covariance about how fast species i changes with respect to changes in the abundance of species j.

Equation (4.9) establishes a relationship between the two different approaches to studying communities described in chapter 1. Recall that the use of linear systems of differential equations formed the basis of many of the theoretical arguments made by MacArthur and his colleagues. Associations among species were also studied using patterns of covariation among species. Patterns of covariation in time are described by the matrix $\mathbf{\Sigma}$ in equation (4.9). One technique that was suggested as a way to estimate interaction coefficients was to use multiple regression to estimate linear relationships among species (e.g., Hallet and Pimm 1979). For temporal data, regression coefficients are calculated using temporal covariances among species, and these covariances are the elements of the matrix $\mathbf{\Sigma}$. But as is evident from equation (4.9), the community matrix cannot be estimated from the elements of $\mathbf{\Sigma}$; rather, it is related to the elements of $\mathbf{\Sigma}^{-1}$, which can be very different from the corresponding elements of the variance-covariance matrix. Thus, regression coefficients from temporal data cannot be used as estimates of the elements of the community matrix, **A**.

An unresolved problem that needs to be mentioned is the use of correlations among species across space as estimates of the elements of the community matrix. Hallet and Pimm (1979), for example, used samples from different locations and assumed that changes from one place to the next represent points along an isocline relating the densities of two competing species. The slope of this line is taken as an estimate of the element of the community matrix relating densities of two species. This approach, however, confounds space and time. Isoclines are explicitly defined as transitory states of a system evolving toward equilibrium. Using samples from different places requires that both the carrying capacity and functional dependencies are the same from one place to the next, and that each censused site represents a transient relationship between the densities of the species as each population is moving toward some global optimum. A similar problem may occur with path analysis (Wootton 1994; Smith, Brown, and Valone 1997).

How Well Does the Linear Model of Lotka Work?

Equipped with a proper method of estimating the parameters of Lotka's kinetic model of community dynamics, we are prepared to examine its effectiveness in describing the dynamics of a real community. Nikkala Pack and I obtained data on a community of seed-eating rodents in the Chihuahuan Desert of Arizona from James Brown. Brown and his colleagues previously showed that, in this community, the removal of competitively dominant kangaroo rats (*Dipodomys* spp.) resulted in an increase in abundances and species richness of smaller species of granivores (Brown and Munger 1985; Brown et al. 1986; Valone and Brown 1995). Removal of these large rodents also had a profound effect on plant communities (Brown and Heske 1990). Thus, we had a data set where the effects of species on one another were known experimentally. The question is whether or not Lotka's equations are capable of describing the results of these experiments with any reliability.

To examine the utility of Lotka's model, we did two separate analyses (Pack 1996). In the first, we used data from two control plots where no rodent removals occurred to fit the linear model to time series of the summed density of large seed-eating kangaroo rats (three species of *Dipodomys*) and the summed density of small seed-eating rodents (species in the genera *Perognathus, Peromyscus,* and *Reithrodontomys*). We summed these densities for two reasons: (1) the data were sparse enough for some individual species that we could not obtain good estimates of the linear model parameters for each species alone; and (2) the major effect of removing the large rodents was an increase in the summed density of all small seed-eating rodents (Brown et al. 1986). The second analysis used community matrices calculated from the control plot to predict the densities of small rodents on two *Dipodomys* removal plots. We used densities summed across two replicated plots for each of these analyses.

Fitting Time Series Data to the Linear Model

The time series for large and small rodents on the control plots is given in figure 4.2. Notice that both series fluctuate considerably over time. Obviously, a linear model cannot fit the entire time series adequately. The linear model cannot produce persistent, irregular oscillations such as those in figure 4.2. It can only produce periodic, convergent, or divergent oscillations. Nevertheless, because competition experiments seem to work over short time scales, there might be a certain period of time over which the linear model provides an adequate description of the dynamics of each species. To evaluate this possibility, we fit the linear model to the time series for large and small rodents by calculating 2×2 community matrices using equation (4.9). Temporal variances and covariances were estimated for both time series using standard formulas, and rates of change were calculated using equation (4.8). Predicted values for the time series were generated using the eigenvalues obtained from the esti-

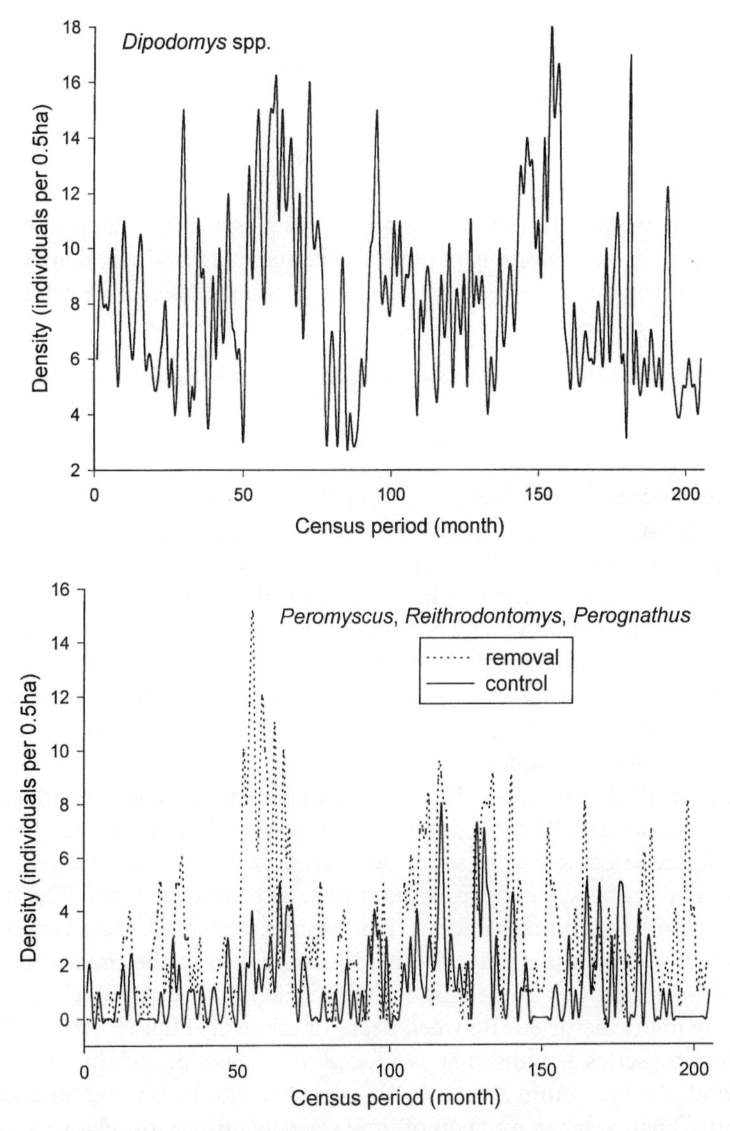

Fig. 4.2 *A,* fluctuations of the combined density of three *Dipodomys* species for 17 years of monthly censuses on the control plots at the Portal site. *B,* fluctuations of the combined density of several species of small rodents for 17 years of monthly censuses at the Portal site on the control and *Dipodomys* removal plots.

mated community matrix. Two types of regression variables were calculated using the eigenvalues of the community matrix. If the eigenvalues were real, then the regression variables were

$$x_j = e^{\lambda_j t},$$

where t is the month in the time series and λ_j is the jth eigenvalue ($j =$ 1,2). If there was a conjugate pair of imaginary eigenvalues, that is, $\lambda_j =$ $a \pm b\,i$ (recall that $i = \sqrt{-1}$), then the regression variables were calculated as

$$x_1 = e^{at}\cos(bt),$$

$$x_2 = e^{at}\sin(bt).$$

A linear regression of the x's on density was then done to produce the predicted time series. In order to examine the possibility that there was a time scale over which these predicted densities best fit the time series, we calculated community matrices and produced predicted values for the time series over a variety of time scales. This was done by taking the first year of the time series, estimating the eigenvalues of the community matrix, and producing an appropriate regression equation, then iterating this procedure by adding an additional month each time until the entire time series was included.

A plot of R^2 goodness-of-fit statistics against the length of the time series used to calculate the community matrix indicated that the best fit was obtained using a time series between 80 and 100 months long (fig. 4.3). The result was similar for both large and small rodents. There are three points worth noting about these results. First, there is a great deal of fluctuation in the ability of the linear model to fit the time series for both species groups as the size of the time series used to calculate community matrices varied. Between peaks of relatively high R^2 values, there were time series lengths that produced very poor fits to the densities. Second, the maximum R^2 achieved decreased with increasing time series length. That is, when a region of time series lengths with relatively good fit was found, the next region with relatively high R^2 had lower maximum R^2 values. This implies that the ability of the linear model to produce a good-fitting model declined with increasing time series length. Third, even for the best time series length (around 75 months), the R^2 was relatively low (around 0.4), indicating that even with the best-fitting model, much of the fluctuation of the time series was not modeled adequately. These observations have some important implications for how we should view the linear community model, but before discussing them, let us look

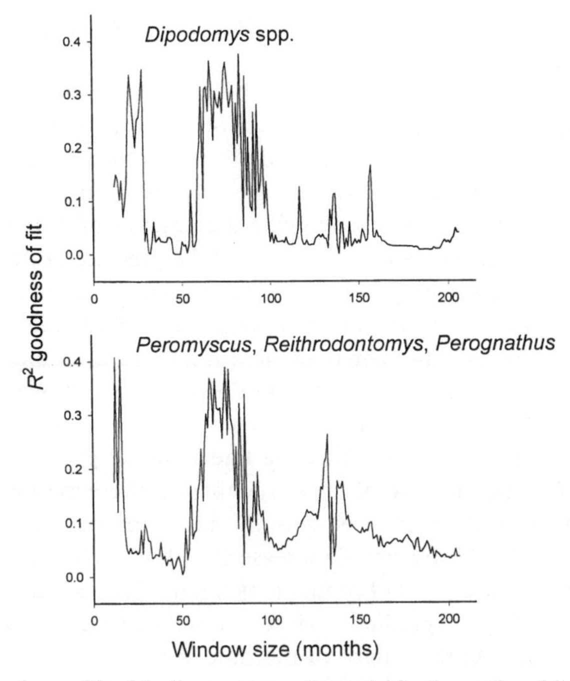

Fig. 4.3 Goodness of fit of the linear community model for time series of different lengths for combined densities of *Dipodomys* species (*top*) and small rodents (*bottom*) for 17 years of monthly censuses on the control plots at the Portal site.

at what results are obtained when the model is used to predict the outcome of an experiment.

Using the Linear Model to Predict the Outcome of a Removal Experiment

Community matrices can be used to predict the outcome of removal experiments such as those conducted by Brown and his colleagues (see, e.g., Yodzis 1989). Suppose that a field experiment changes the density of species j by the amount ΔN_j (this could be an increase or decrease), while not changing the density of any other species. If this is a regular change (e.g., a press experiment, Bender, Case, and Gilpin 1984), then the change in the equilibrium density of species i will be

$$\frac{\partial N_i^*}{\partial \Delta N_j} = -a_{ij}^*, \qquad \textbf{(4.10)}$$

where a_{ij}^* is the i,jth element of the inverse of the community matrix $\mathbf{A}^{-1}$, and N_j^* is the equilibrium density of species i. Note that if the press perturbation is a removal, then $\Delta N_j < 0$, and the negative sign in equation (4.10) disappears.

Equation (4.10) was used to predict the difference between the abundance of small rodents on the control plot and on the plots where all large seed-eating rodents were removed. Note that the predicted quantity is the difference between equilibrium density (N_i^*) on the control plot versus the experimental plot. Following the practice of estimating equilibrium densities with means, we compared the predicted difference that we obtained by subtracting the mean density on the control plot from the mean density on the experimental plots to that predicted by equation (4.10). As before, we thought that there might be a temporal scale at which the linear model might provide the best predictions, so we repeated the calculations for time windows of different lengths.

The time window within which the linear model produced the best predictions for the increase of small rodents on the experimental plots was between 37 and 45 months, with the best time scale of 39 months being within 1–2% of the actual increase (fig. 4.4). Before and after this time window, the accuracy of predictions fluctuated drastically, with periods of extremely poor predictions interspersed with periods of relatively good predictions. As the length of the time series used to calculate the

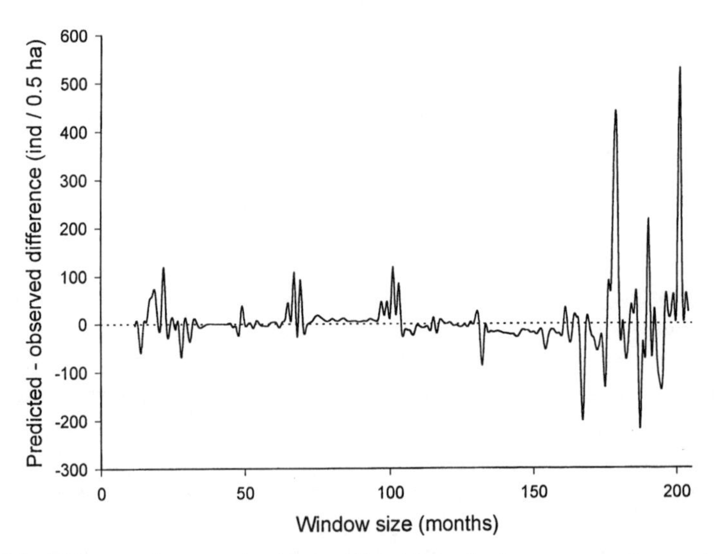

Fig. 4.4 Differences between densities predicted from the linear community model and observed densities of small rodents for 17 years of monthly censuses on the control plots at the Portal site. Note that the predictions are best at around 37 to 45 months, and that accuracy varies considerably for time series shorter or longer than this "characteristic" scale.

community matrix for the control time series increased, the ability of the linear model to produce reliable predictions decreased, even during periods of time for which the model produced relatively stable results (fig. 4.4). These results imply an extreme sensitivity of the linear model to the particular time series used to calculate its parameters.

Given the sensitivity of the linear model's predictions to particular time series, we investigated the range of this behavior by varying the particular time series that we used in our calculations. This was accomplished by starting with the first point in the time series, generating predictions using all censuses conducted after that point, finding the best time scale (for this time series it was 39 months), then repeating the procedure 100 times, each time deleting the leading observation. Thus, we had 100 time series, the first 205 months long, the second 204 months long, and so on until the last was 105 months long. This gave us 100 estimates of the "characteristic" time scale, that is, the time scale for which the model best predicted the actual results of the removal experiment. The distribution of these characteristic scales was highly right skewed (fig. 4.5). The modal scale was about 2 years; that is, the most common period of time over which the linear model produced its best predictions was about 2 years, with a few time series having characteristic scales much longer (the longest scale was 15 years).

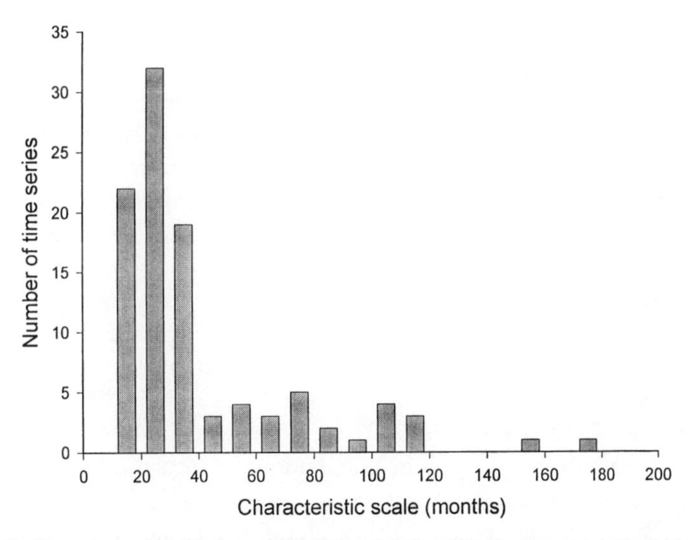

Fig. 4.5 Frequency distribution of "characteristic" scales for linear community model predictions from time series of different lengths. The characteristic scale is the length of time at which the best prediction was obtained from the linear community model.

Limitations of a Linear Model in a Nonlinear World

From the example analyses in the previous two sections, some common themes emerge, which I examine in this section. But it must be kept in mind during this discussion that the results obtained for a relatively clear-cut, simple experimental system may be unique to the system. How much can the specifics obtained from the desert rodent community be applied to other systems and how much of these results are unique? I cannot answer these questions directly, since similar analyses have not been done for other systems. There is nothing to make me think, however, that this data set is atypical or unusual, so I suspect that similar kinds of results would be obtained for other systems, though they certainly would differ in the quantitative details. I will discuss here qualitative conclusions that I suspect will be fairly general.

It is clear that the dynamics of the time series examined in the previous sections are much more complicated than the solution to a system of linear differential equations can model. The dynamics that can be produced by such a solution are limited to simple approaches to equilibria, converging or diverging oscillations, or simple periodicity. The implication is that in the equations describing population dynamics (eq. [4.1]), the functions f_i that describe per capita rates of growth for each species must contain significant nonlinearities, substantial amounts of stochasticity, or both. Nonlinearities may arise when these functions contain multiples and powers of densities of each species, or when there are significant time lags in the effects of species on themselves or other species. For example, a competitor may reduce the rate of reproduction of another species, but it may take several months for this reduction to change the density of the affected species after juveniles are recruited into the population. Stochasticity arises from either environmental or demographic uncertainty. Environmental uncertainty is due to irregular fluctuations in food, habitat quality, or climate that affect population processes. Demographic uncertainty results from the fact that the processes underlying population change are carried out by individual organisms that must be in specific places at specific times in order to interact with conspecifics.

Note that there is a subtle difference between the linear model described by equation (4.3) and the conventional formulation of the Lotka-Volterra competition equations given in equation (4.6). To obtain a truly linear system, N_i was replaced by its equilibrium value to get equation (4.3). In equation (4.6), N_i is included instead of N_i^* in the equation for each species, so there are actually nonlinear terms in the equations (e.g., N_i^2 and N_iN_j). It turns out that when there are three or more species in such a system (i.e., $s > 2$), there is the possibility for complicated dynamics in the system. Yodzis (1989) discussed a variant of equation (4.6) that models the relationship of a predator and two prey species. This version of the model can generate rather complicated dynamics when parameters are chosen appropriately. In the next chapter I consider some of the challenges posed by studying such systems.

When the linear model was fit to time series, the ability of the model to produce an acceptable statistical fit declined as more of the time series was included in the calculations. It performed best at a relatively short time scale of about 5 years. Data sets longer than this were increasingly unable to be fit by the linear model. The same thing happened when the linear model was used to predict the outcome of the removal experiment. The time over which accurate prediction was possible was relatively short (2–3 years), after which the ability of the linear model to predict deteriorated until it was essentially nonexistent. The ability to predict systems governed by strange attractors deteriorates with time, although this is not quite the same thing as we have encountered with the rodent data sets. Here we used a specific kind of model that makes certain assumptions about the existence of equilibria. What we found is that this model became more inadequate the longer we observed the system.

A final point should be made regarding the linear model. Recall that the model's ability to fit time series and predict the results of removal experiments fluctuated considerably depending on the particular length of the time series used. This may be partially an artifact of the mathematical structure of the model. Extreme values can greatly affect the outcome of calculations on matrices involving inverses and eigenvalues. Since the original time series often fluctuate considerably from one month to the next (fig. 4.2), it is likely that adding additional observations often produced relatively unstable estimates of eigenvalues and inverses, and so was responsible for the fluctuations in model performance (figs. 4.3 and 4.4). Although there is some measurement error that might go into the determination of population densities, it is unlikely that the fluctuations observed are completely random (see Brown and Zeng 1989). If the nature of the mechanisms that cause population dynamics in the community introduces such fluctuations, then the problem is with the mathematical model and not the data.

The Limits of Linear Community Theory

Linear community theory can be defined as that body of ecological and evolutionary theory that uses community matrices and Lotka-Volterra-like descriptions of communities to derive theoretical predictions regarding community structure and dynamics. Such theory is typified by the works of MacArthur (1972) and his colleagues, but is in no way limited to this kind of research. Linear thinking permeates much of experimental ecology as well. The expectation of straightforward results from experiments that perturb natural ecosystems is very clearly based on linear thinking.

One of the purposes of going through the rather detailed analyses of the desert rodent system in this chapter was to introduce to the reader the distinct possibility that linearity is a temporary way of getting at the complexity of biological communities. Although much of the fundamental structure of the rodent community could be obtained from the analysis, the value of the description decayed with time. This is probably true of almost any prediction or system description one could make about an ecological system. As I pointed out near the beginning of this chapter, a number of analyses of different communities using community matrices will be necessary before we can decide exactly what the limitations of the method are. But my suspicion is that what we have observed for the rodent system will be typical. The reason for this suspicion can be found in chapter 3. Recall that there was a clear time effect in the meta-analysis of competition performed by Gurevitch et al. (1992; see also chap. 3 in this volume, fig. 3.2). I predict that community matrix analysis will come to be seen as scale dependent, only adequate to predict over relatively short time scales.

What does this say about theories built using community matrices and associated constructs? I tentatively conclude that predictions derived from such theories will only be valid over relatively short time and small spatial scales. The implications of the rodent analyses for short time scale studies is clear; however, their implications for aggregates of populations in space is less apodictic. Aggregating populations in space may reduce error variances or sources of process uncertainty, but may also obscure the very processes being studied. Predictions about community structure, such as assembly rules or niche partitioning, might apply in ecological time, but not over longer time scales. Niche partitioning might occur as an ecological process, but not as an evolutionary one, as implied by early treatments of the problem (MacArthur and Levins 1967; Levins 1968). This does not mean that phenomena such as character displacement cannot occur, it simply means that if differences between species evolve as a consequence of competition, this evolution must occur on much larger spatial scales than that typified by studies of local communities. Experimental demonstrations of competition within local communities do not logically lead to the conclusion that selection has occurred to reduce ecological similarities among species.

Results of small spatial scale studies of communities, at least using current techniques and theories, are scale dependent. We cannot simply assume that what we demonstrate at one location or point in time is generalizable to other locations or other times. When we examine communities on small scales, there is indeed complexity; and though we can learn something from this complexity through experimental and theoretical investigations using "traditional" methods based on linear thinking, what we learn is limited to small spatial and short time scales. We see complexity, in part, because we see only part of the system through a conceptually limited window. As many authors have stated, scaling up in space and time is much more difficult than simply putting together independent pieces of information from different systems (Wiens 1989b; Allen and Hoekstra 1992; Levin 1992; Turner, Gardner, and O'Neill 1995).

It is possible that the mechanisms known to be involved in determining structure lead to more complicated theoretical structures. Those mechanisms undoubtedly involve time lags and other processes that ensure that equations describing communities on small scales do indeed contain many nonlinear terms. So the limitations of the linear approach may not apply to approaches that assume that nonlinearities dominate the kinetics of community change. Generalizations may yet emerge if we use nonlinear mathematics to generate theories about communities. Recall that in chapter 2 I argued that "chaos" is not really unstructured, as some people conventionally interpret it to be. Strange attractors have a stability at larger scales that appears to be indeterminate (and indeed *is* indeterminate in one sense) at smaller scales. This possibility leads us to the next chapter, where I examine what a nonlinear approach to local community dynamics might look like.

Appendix 4A: Estimating the Average Per Capita Rate of Population Change over a Finite Time Interval

Here I consider how to calculate average rates of change of a population between census intervals. Suppose we census a population every Δt time units, say, once every year or month. On the census conducted at time t the population density of species i is $N_i(t)$, and on the next census it is $N_i(t + \Delta t)$. We assume that between t and $t + \Delta t$ the change in population density is governed by equation (4.1). The average per capita rate of change over the time interval (R_i), then, will be

$$R_i = \frac{\int_0^{\Delta t} f_i \, dt}{\Delta t}, \qquad \textbf{(4A.1)}$$

where f_i is the function describing the effects of all species in the community on the per capita rate of change for species i.

Although we do not know what values f_i takes on, we can approximate its behavior during the time interval Δt by replacing it with the average per capita rate of change. Substituting R_i for f_i in equation (4.1) gives

$$\frac{1}{N_i}\frac{dN_i}{dt} = R_i. \quad \textbf{(4A.2)}$$

Since R_i is an average, and by the definition given in equation (4A.1) it is constant, the solution to equation (4A.2) is easily obtained as

$$N_i(t + \Delta t) = N_i(t)\, e^{R_i \Delta t}.$$

Rearranging this equation then gives the estimate for R_i as

$$R_i = \frac{\ln N_i(t + \Delta t) - \ln N_i(t)}{\Delta t}. \quad \textbf{(4A.3)}$$

It is also interesting to note that as Δt approaches zero, equation (4A.3) R_i approaches f_i. In fact, calculus textbooks use the limit of equation (4A.3) as a formal definition of a derivative, which is what f_i is.

Appendix 4B: Estimating the Elements of the Community Matrix A

Recall that the dynamics of a species can be described by the general equation

$$\frac{dN_i}{dt} = N_i f_i(N_1, N_2, \ldots, N_s).$$

Now define the quantity $R_i(t) = (1/N_i)(dN_i/dt)$. Expanding this equation about average densities, which we take as our estimates of equilbria over some finite time interval, and ignoring higher-order terms, we have

$$R_i(t) = \sum_{j=1}^{s} \delta_{ij}(N_j - \overline{N}_j), \quad \textbf{(4B.1)}$$

where $\overline{N}_j$ is the average density of species j, and δ_{ij} is the partial derivative of f_i with respect to N_j evaluated at mean densities. By making appropriate substitutions, this approximate solution is equivalent to the conventional Lotka-Volterra equations.

Given this expansion of the general equations of population dynamics about average densities, it is possible to describe the relationship between *functional dependency*, represented by the partial derivatives of the per capita rate of population change functions, and *statistical dependency,* or the correlation of population densities of species with each other. Equation 4B.1 can be rewritten as a matrix equation:

$$\mathbf{R}(t) = \mathbf{D}\,[\mathbf{N}(t) - \mathbf{N}], \qquad \textbf{(4B.2)}$$

where $\mathbf{R}(t)$ is a column vector of rates of change for each species at time t, $\mathbf{N}(t)$ is a column vector of population densities at time t, $\mathbf{N}$ is a column vector of average densities, and $\mathbf{D}$ is a matrix of interaction coefficients, δ_{ij}. Rearranging gives

$$[\mathbf{N}(t) - \mathbf{N}] = \mathbf{D}^{-1}\,\mathbf{R}(t).$$

Postmultiplying both sides by $[\mathbf{N}(t) - \mathbf{N}]'$ gives

$$\mathbf{S}(t) = \mathbf{D}^{-1}\,\mathbf{R}(t)\,[\mathbf{N}(t) - \mathbf{N}]'.$$

The $s \times s$ matrix $\mathbf{S}(t) = [\mathbf{N}(t) - \mathbf{N}]\,[\mathbf{N}(t) - \mathbf{N}]'$ contains the squared deviations of species densities away from their means along the diagonal, and the off-diagonal elements contain the cross products of deviations from means for each pair of species. To find the covariance matrix, $\boldsymbol{\Sigma}$, it is necessary to average $\mathbf{S}(t)$ over the time interval for which the community is stationary. Letting $t = 0$ represent the start of the interval and $t = T$ the end of the interval, the covariance matrix of species densities is given as

$$\begin{aligned}\boldsymbol{\Sigma} &= \frac{\int_0^T \mathbf{S}(t)\,dt}{T}\\ &= \mathbf{D}^{-1}\,\frac{\int_0^T \mathbf{R}(t)\,[\mathbf{N}(t) - \overline{\mathbf{N}}]\,dt}{T} \qquad \textbf{(4B.3)}\\ &= \mathbf{D}^{-1}\,\mathbf{Z}.\end{aligned}$$

The elements of the $s \times s$ matrix $\mathbf{Z}$ are given as

$$z_{ij} = \frac{\int_0^T R_i(t)\,[N_j(t) - \overline{N}_j]\,dt}{T}.$$

Thus, the elements of the matrix $\mathbf{Z}$ are the average cross product of the rate of change of species i times the deviation of species j from its average density. From equation 4B.3, it is possible to estimate the matrix of interaction coefficients as

$$\mathbf{D} = \mathbf{Z}\,\boldsymbol{\Sigma}^{-1}. \qquad \textbf{(4B.4)}$$

The matrix **D** contains the magnitude of the functional dependencies of species in the community. The matrix **Σ** contains a description of the statistical dependencies of species on one another. Notice that statistical dependence between two species does not imply functional dependence. If the appropriate elements of the matrix **Z** are zero, then there could be statistically significant correlations among the densities of the species without any functional dependence because the corresponding elements of **D** would be zero. Hence, estimates of species interactions based on covariances (e.g., Schluter 1984; McCulloch 1985) or regression (e.g., Crowell and Pimm 1976; Hallet and Pimm 1979) do not necessarily contain information regarding functional dependencies among species expressed by the elements of the matrix **D**.

Given estimates of the coefficients in equation 4B.4, it is possible to obtain a solution to the system of equations under the assumption of linear interactions among species. As before, write the system of equations as

$$\mathbf{d}(\mathbf{N} - \overline{\mathbf{N}}) = \mathbf{diag}(\overline{\mathbf{N}})\ \mathbf{D}\ (\mathbf{N} - \overline{\mathbf{N}}). \qquad \textbf{(4B.5)}$$

Then the solution to equation 4B.5 for the ith species is

$$N_i = \overline{N}_i + \sum_{j=1}^{s} c_{ij}\, e^{\lambda_j t}, \qquad \textbf{(4B.6)}$$

where λ_j is the jth eigenvalue of the community matrix $\mathbf{A} = \mathbf{diag}(\overline{\mathbf{N}})\ \mathbf{D}$. If there are significant nonlinearities in the interactions among species, then equation 4B.6 is expected to produce a relatively poor fit to the actual time series of species densities in the community. However, the eigenvalues could be used in theory to obtain estimates of the Lyapunov exponents of the system of equations governing community dynamics and, hence, to obtain some idea of the dimensionality of the community. However, there are better ways to estimate Lyapunov exponents that are more general and reliable (Ellner and Turchin 1995).

CHAPTER FIVE

Communities as Nonlinear Systems

> You remember the story of the wise man who invented the game of chess. As a reward, he asked the king to put one grain of rice on the first square of the chessboard, two grains on the second square, four on the third, and so on, doubling the number of grains of rice at each square. The king first thought that this was a very modest reward, until he found that the amount of rice needed was so huge that neither he nor any other king in the world could provide it.
>
> D. Ruelle (1991)

Certainly one of the most remarkable properties of organisms is their ability to reproduce themselves. In fact, individual organisms usually produce as many offspring as is feasible within the constraints imposed by their unique life history. This tendency of individual organisms to reproduce as many offspring as possible means that a population of organisms could potentially increase without bound. This is the fundamental insight of Malthus that Darwin borrowed to build his theory of natural selection. The Malthusian principle is as close to an ecological law as we have. The tremendous potential of a population to increase is due to positive feedback mechanisms, which tend to magnify the consequences of small effects.

But, of course, biological populations do not increase without bound. There are many "forces" operating in nature that prevent such increase. As the founder of wildlife ecology, Aldo Leopold, so eloquently stated: "We may conceive of a population as a flexible curved steel spring, which, by its inherent force of natural increase, is constantly striving (so to speak) to bend upward toward the theoretical maximum, but which the various factors are at the same time constantly striving to pull down" (1933, 24). To Leopold, and to most ecologists, population dynamics are the outcome of tension between reproduction and the limiting effects of the environment. Reproduction pushes the population upward, while limited resources, predation, and disease pull the population downward. Biological populations can be said to be "driven" and "damped": driven by the boundless ability of organisms to reproduce and damped by the unceasing power of nature to destroy organisms.

Physical systems that are simultaneously damped and driven rarely have simple kinetics. The study of chaotic systems in general has made much progress from consideration of damped and driven physical systems (Peitgen, Jürgens, and Saupe 1992). One of the first chaotic systems discovered was found in an attempt to explain the movement of air masses as a simplified hydrodynamic process (Jackson 1990). In such a process, energy is supplied to a viscous fluid by a temperature gradient, which provides the driving force for the hydrodynamic flow. Gravity, the viscosity of the fluid, and other forces constrain the movement of the fluid, essentially using up energy to "slow" its motion. For the earth's atmosphere, the temperature gradient is provided by the sun, while gravity and the physical properties of air provide constraints. Lorenz (1963) greatly simplified the mathematics describing the hydrodynamics of the earth's atmosphere to obtain a system of three nonlinear differential equations. He found that when he estimated values for the parameters of these equations for the atmosphere, the resulting dynamics were chaotic. The solution to Lorenz's equations is illustrated in figure 2.3. The point I emphasize here is that the atmospheric model used by Lorenz described a damped and driven system.

Biological populations, in being damped by the environment and driven by reproduction, easily fall within the class of physical systems that *could* exhibit chaotic dynamics. Yet despite this expectation, which is based on the nature of how populations operate, there has been relatively little support among ecologists in the past for the idea that population dynamics might generally be chaotic. Most ecologists have been skeptical of the complex statistical techniques necessary to detect chaotic systems, and they have been equally hesitant to embrace the complex behavior that such systems exhibit. There is sufficient noise in ecological measurements and enough variability in environmental conditions that many ecologists seem to think that a relatively simple linear model of dynamics over time (with any variation being ascribed to measurement error, "the natural variability of the system," or both) is as good at explaining the dynamics of a population as any hideously complex set of differential equations possessing bizarre mathematical properties might be.[19] There is good reason for such skepticism. Over the years, many ideas have been borrowed from systems science and applied to ecosystem

19. This kind of thinking seems to persist despite the problems with linear statistical models discussed in chapter 2, and the limitation of linear dynamical models encountered in chapters 3 and 4.

and community analysis. Most of the time these ideas have been viewed as a passing fad. So why expect chaos to provide any new insights?

Skepticism regarding chaos has been heightened by the inability of biologists to detect widespread existence of chaotic population dynamics. One of the first empirical attempts to examine the role of chaos in population dynamics used a relatively simple population model to analyze the dynamics of a number of insect populations (Hassell, Lawton, and May 1976). Only 1 population out of 28 had parameter values that would generate chaotic dynamics. Furthermore, techniques that were initially developed to identify chaos in time series required far more data than were available for ecological systems. So even if communities were chaotic, it was initially impossible to determine that fact because there simply was not enough data.

This situation changed as new techniques were developed to identify the existence of chaos in shorter time series that include measurement error (e.g., Sugihara and May 1990a,b; Sugihara, Grenfell and May 1990; Ellner and Turchin 1995; Barahona and Poon 1996). With these new techniques, it became possible to look for evidence of the strange attractors that generate chaotic dynamics. Surprisingly, there have been relatively few examples of chaotic systems that have arisen from examination of ecological time series (Turchin and Taylor 1992; Ellner and Turchin 1995), but this may be due partly to the relatively conservative nature of the tests used (see, e.g., Ellner and Turchin 1995). In this chapter, I outline how the description and analysis of local communities might proceed if they are dominated by nonlinear processes that produce complicated dynamics. As with the linear approach, there are strengths and limitations inherent in a nonlinear approach to community structure. It is unclear whether the generalities sought by community ecologists will emerge from a nonlinear community ecology. I suspect, as I argued in chapter 3, that ultimately local communities are too small to exhibit the kind of structure from which generalities could emerge. But this remains to be seen.

Nonlinear Community Dynamics

If communities are dominated by processes that cannot be modeled adequately by a system of linear equations, then how are we to describe their dynamics? There are at least two general classes of effects that might compromise the ability of linear equations to describe community dynamics. First, there may be stochastic processes that operate within the community to perturb the system away from dynamics expected from linearity. Second, the deterministic component of community dynamics might involve processes that operate multiplicatively. Density dependence and slowly saturating functional responses are two processes that might add nonlinearities to community dynamics. These two possibilities are not mutually exclusive. In some systems, small effects introduced through stochastic processes might be magnified by nonlinearities and profoundly affect the future dynamics of the system. At least some chaotic systems, however, can be described using macroscopic quantities that are related to the deterministic processes that underlie the system's behavior. Ergodic chaotic systems can be considered stable from a macroscopic perspective. If biological communities can be modeled by systems of equations that describe ergodic chaotic systems, then the problem is to develop appropriate macroscopic measures of a community that capture the essence of its dynamics.

Chaotic systems described by low-dimensional chaotic attractors are fairly amenable to empirical analysis. A low-dimensional attractor occurs when a few fundamental processes drive the dynamics of the system being studied, even when a relatively large number of objects is involved. The question of interest here is whether community dynamics are controlled by a few fundamental processes or by many. On the one hand, if community dynamics are primarily determined by biotic interactions among species, then it might be expected that a few general "forces," such as competition or predation, are responsible for fluctuations in abundance among the species involved. On the other hand, since communities are open systems, there are many influences on community dynamics—such as weather, migration, variability in competitive and predation effects, and so forth—that could each have a very important effect on what happens to the community over time. So it is not entirely clear that communities can be modeled adequately by low-dimensional chaos.

There are also chaotic systems that are not ergodic. That is, they do not have a stable attractor that remains unchanging over time (e.g., Sommerer and Ott 1993). In such a system, the statistical distribution describing the states of the system over time changes. That is, the mean, variance, and other statistical properties that describe the system are functions of time. It is unclear whether such systems can be described macroscopically. Perhaps the statistical distribution of estimated means or variances might indicate some higher-level macroscopic structure, but this is unclear. Furthermore, it seems reasonable to expect that the more complicated the system described by a set of nonlinear equations is, the more likely it is that those equations describe a nonergodic chaotic system. Open communities might be complicated enough to ensure that their kinetics must be described by nonergodic dynamical equations.

Based on considerations outlined in the previous two paragraphs, it might be argued that the description of communities from a nonlinear perspective will run into limitations similar in nature to those encountered in chapter 4 with the mathematically simpler linear systems approach. This is really an empirical issue. I will defer discussion of it until later, because I want to consider first the case where it is assumed that community dynamics are amenable to analysis as ergodic, low-dimensional nonlinear systems.

Communities as Ergodic Systems

Chaotic systems that are ergodic have some very useful properties. One of the most important of these is the following: The longer one observes a part of the system, the more information about the entire system one can obtain. This property is associated with the macroscopic stability of the system. If the shape and nature of the attractor governing the system remains unchanged, the system will undergo dynamics such that certain combinations of values for the variables describing the system are rarely or never obtained. This shows up in statistical distributions of values obtained from the time series, as we discovered in chapter 2. However, if the attractor that determines system dynamics changed in a fundamental way over time, then no information on the nature of the attractor could be obtained statistically, because statistical quantities calculated from such a system would themselves change over time.

If one knows the underlying dynamical equations that govern the system being studied, then the statistical stability of the solution is an interesting property, but the mechanics of the system are known because the equations have been formulated from those mechanics. Unfortunately, in many applications, particularly those involving populations and communities, the equations describing system dynamics are unknown. Recall the problem that Hassell, Lawton, and May (1976) had when describing insect populations. They assumed that population dynamics of insects could be described by a relatively simple equation, and their results depended heavily on this assumption. One might come up with different equations and obtain different results. To avoid such assumption-specific problems, one must admit that the equations governing the system being studied are unknown.

The stability inherent in the macroscopic structure of an ergodic system allows the development of techniques for recovering some information about the original system. A number of rather complex statistical methods exist for recovering such information. Although many techniques require large data sets (thousands of data points), recently developed techniques can be used with much smaller samples. Usually, these techniques still require sizable samples; usually in excess of 100 data points are needed (e.g., a time series of monthly censuses conducted across 8 years). Many such time series exist in the literature. In the next section, I consider how one might go about analyzing the dynamics of ecological systems that are assumed to be stable in a statistical sense (i.e., ergodic).

Ergodicity of community dynamics can be assessed by examining the averages over time of population densities and the temporal autocorrelation structures of the constituent species (Royama 1992). The basic result is that if the average autocorrelation between densities at different times vanishes with increasing time, then the sample statistics converge on stable values. Consider the data from the Chihuahuan Desert ecosystem that I discussed in the last chapter. For the *Dipodomys merriami* time series shown in figure 4.1B, it is not clear that the average autocorrelation converges on zero (fig. 5.1). Note that for lags of up to 12 months, there is a positive but declining autocorrelation. That is, for the next 12 months, on average, after a given census, censuses will tend to be similar to previous ones. Censuses separated by a period of 1–3 years were negatively correlated. Censuses separated by 3–4 years were not correlated, but those between 4 and 5 years were positively correlated. Censuses separated by more than 5½ years were negatively correlated. The average correlation dropped rapidly until a lag of about 4 years, after which it

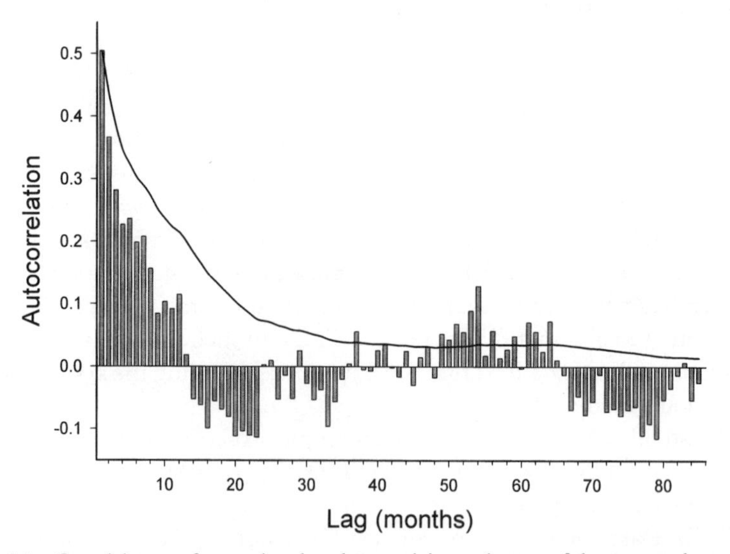

Fig. 5.1 Correlelogram for species abundances giving estimates of the temporal autocorrelation between censuses as a function of the time separating them. The solid line is the cumulative average autocorrelation. Note that the average autocorrelation declines over time, but at a decreasing rate.

dropped very slowly toward zero. Estimates of autocorrelations between censuses conducted more than 90 months apart appeared to be unreliable since they were many times smaller than their standard errors. It appears, however, that the *D. merriami* time series might be approximately ergodic. At least some populations might approach ergodicity, particularly in the absence of any trends in the data.

Modeling Ergodic Ecological Systems

We saw in chapter 4 that Lotka's original formulation implied that near equilibrium we could model a community with a system of linear equations. The limitations of this approach center around the scale at which one wishes to describe the system. If the goals of the study are short term, or limited to a small geographic region, then linear analysis of the community might be entirely appropriate. However, if such limitations are unacceptable for a particular situation, how might we proceed to analyze data on community dynamics? As mentioned in the previous section, the macroscopic stability of an ergodic chaotic system has allowed the development of techniques for characterizing the dynamics of such a system, so we might inquire whether such techniques might be used to analyze local communities. This requires that we assume that communities can be described adequately as low-dimensional chaotic attractors.

In an ergodic chaotic system, the nature of the attractor in state space[20] is such that the system regularly visits specific regions of that space. Each time the system revisits a specific region, it provides some information about the nature of the system. The kind of stability evident in statistical distributions that I described in chapter 2 applies to the chaotic attractor. Since the system is continually revisiting certain regions of the state space, the longer the system is observed, the more information is recovered regarding the system.

Techniques for recovering information about an unknown system from a time series of one of the state variables describing that system are based on comparing current values of the time series to values at earlier times. Since a community that is governed by a strange attractor is constrained to revisit regions of state space time after time, the equation for a single species in that community can be written as

$$\frac{dN_i}{dt} = N_i g_i[N_i(t), N_i(t - \tau_1), N_i(t - \tau_2), \ldots, N_i(t - \tau_p)], \quad \textbf{(5.1)}$$

20. State space refers to the set of variables that is used to describe the system. In a biological community, the state space would consist of variables that describe the density of each species. Every census would represent a point in this state space, and a series of censuses would correspond to a collection of points.

where g_i is a function that relates the per capita rate of population change to the density of the species in the past. The τ's indicate that the per capita rate of change at the current instant in time (t) is related to the abundance of the species at a set of times in the past. The τ's need not be equally spaced, although conventional time series analysis usually assumes that they are (e.g., Wei 1990; Chatfield 1996). The assumption implicit in this formulation is that if the community is governed by an ergodic nonlinear set of equations, then the dynamics generated by the system are "information preserving," in the sense that over a certain period of time it is possible to distinguish trajectories that were initially different (Shaw 1981). Hence, a description regarding the entire system is often recoverable from a sufficient number of observations of a single component of the system. This assumption is not a bad one to make as long as the processes generating changes in the system ensure ergodicity.

Note that this approach to describing a community implies an equivalence between the description of community kinetics given by equation (4.1) and the time-lagged description of each species given by equation (5.1). Such an equivalence, however, does not mean that by observing the dynamics of a species over a sufficiently long period of time we can reconstruct the processes that caused those dynamics. For example, we could not necessarily determine if a species' dynamics had been caused by the effects of one or more of its competitors simply by observing its density over a long period of time.

Hence, despite the availability of statistical techniques to analyze communities that might be governed by nonlinear processes (and hence, sets of nonlinear equations), techniques for studying chaotic systems will be of limited use in determining cause-effect relationships within local communities from observational studies. These limitations are not unlike those that have constrained the use of ideas derived from the linear community theory to observational studies (e.g., Wiens 1989a,b). So shifting ecology to a predominantly nonlinear paradigm will not necessarily improve what can be done with observational studies.

Although it may be unlikely that observational studies of communities using nonlinear techniques will provide any improvement beyond conventional approaches, the question naturally arises whether these techniques, when applied to experimental systems, might allow ecologists to overcome the limitations imposed by the linear methods described in chapter 4. Recall that in chapter 3 I argued that there are significant limitations on interpretations of experiments conducted on local communities because often the effects of experiments change over time. Can nonlinear approaches improve the situation? The answer depends very much on the kind of experiment being conducted. A "pulse" experiment perturbs the community once by a removal or adjustment in species densities, and tracks the community's response thereafter. A "press" experiment maintains an experimental removal or density adjustment for the life of the experiment (Bender, Case, and Gilpin 1984).

For a pulse experiment, there is no guarantee that after the perturbation the community will return to its preexisting state. If a pulse experiment is followed for only a short period of time after the manipulation, then it is likely that the experimenter will miss this. If nonlinearities exist in the processes governing community dynamics, it is not appropriate for an experimenter to simply assume that after the perturbation, the manipulated community will return to equilibrium. Interpretations of experimental results that make this assumption are invalid. One of the hallmarks of chaotic nonlinear systems is that their dynamics can be quite different when the system is "restarted" in a different region of state space than the one it currently occupies. Hence, pulse experiments will have some rather profound limitations. Clearly, such experiments will require careful pre- and postperturbation data to ensure that the observed effects are not transient in nature and do not shift the community into a different dynamical state.

In a press experiment, a perturbation is continually applied to the community, so that unless there is absolutely no effect of the perturbation, the abundances of species within the community must change in response to it. Even within the same system, different taxa may react very differently to the continual perturbation. In a linear system, a press experiment leads to a new equilibrium (Yodzis 1989), but, as we saw in chapter 4, this may be of limited usefulness in community analysis. In a nonlinear system, a press experiment could change the nature of the strange attractor governing the system's dynamics. Furthermore, there is no guarantee that a nonlinear system that is ergodic before a press perturbation will remain so afterward. So it is not clear that techniques commonly used to study nonlinear systems will allow community ecologists to overcome the kinds of limits inherent in the linear approach.

Are Communities Ergodic?

An important question regarding the use of nonlinear techniques in analyzing communities is whether communities can be considered ergodic. If not, techniques that implicitly or explicitly assume ergodicity cannot be expected to provide adequate descriptions of how a community changes over time or responds to disturbances or other perturbations. Ergodicity implies that the parameters describing the properties of a statistical distribution are constant in space and time. It is often possible to parameterize a distribution in terms of its moments, the first two being the mean and the variance. Hence ergodicity implies that, in a community, the mean and variances of species densities do not change over time. But the stability of the statistical description of a community may itself be scale sensitive. Aggregating data across space and time may change the stability properties of the statistics of the community. If this is true, it may be difficult to reach conclusions about the mechanisms determining structure based on statistical description of a community, even in the best of experiments.

Empirical evidence suggests that it is questionable to assume that population dynamics of species within ecosystems result in ergodic time series (Pimm and Redfearn 1988; Pimm 1991; Curnutt, Pimm, and Maurer 1996). As an example, consider changes in variability within the Chihuahuan Desert rodent community analyzed in chapter 4. The variance appeared to level off over time in the time series for one of the dominant kangaroo rat species, but appeared not to for one of the smaller species (fig. 5.2). Measuring variability for the entire community as a single entity is not straightforward. There are two different measures of variability often used with multivariate data (Rencher 1995). The first

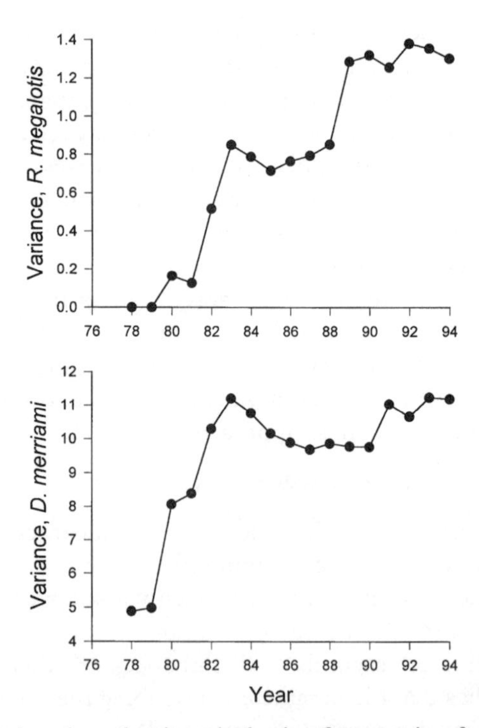

Fig. 5.2 Changes in estimated variance in density of two species of granivorous rodents in a Chihuahuan Desert ecosystem over time. Variances are cumulative, calculated yearly.

is the trace of the covariance matrix, referred to as the *total variance,* which turns out to be the sum of the variances of densities for each individual species. The second is the determinant of the covariance matrix, called the *generalized variance,* which includes information about correlations among variables (in this example, species densities). The two measures of variability for the Chihuahuan rodent community demonstrate different time evolutions (fig. 5.3). The total variance increases rapidly for the first six years of the study, then seems to level off or increase only slightly thereafter. The generalized variance, however, barely increases for the first five years, then begins increasing and continues to do so for the rest of the study period. Thus, although each species seems to reach a relatively stable variance, their correlations with one another continue to evolve. It would be questionable to assume that this particular community is evolving an ergodic probability distribution. In other words, the community as a whole does not seem to be converging on a stable configuration. Looking at the species composition of the community further

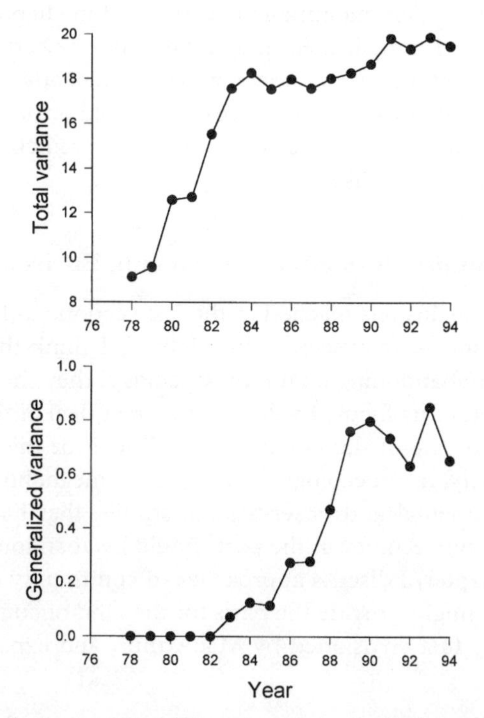

Fig. 5.3 Changes in the variability of the entire rodent community of granivorous rodents in a Chihuahuan Desert ecosystem over time. Both measures of variability are based on multivariate statistical generalizations of the variance. See the text for definitions of each multivariate measure of variability.

drives this point home. One of the dominant species, the banner-tailed kangaroo rat (*Dipodomys spectabilis*) became extinct during 1985, and the community has been invaded by species typical of grasslands (Valone and Brown 1995). These kinds of processes have led to an entire reorganization of the rodent community over the course of the experiment (Brown, Valone, and Curtin 1997). For the Chihuahuan Desert rodents, it seems unlikely that the community, strictly speaking, can be considered ergodic, although the density of its most abundant member, *D. merriami*, might be.

If communities are not ergodic, then the tentative conclusions made in the previous section are further extended. Although the mathematics of ergodic chaotic systems seems in some respects to mimic the kinds of dynamics that we see in local community assemblages, the advantages of using statistical techniques based on the assumption that communities are ergodic are likely to be limited. Such techniques might be used for forecasting,[21] but may not provide a mechanistic understanding of why a community responds the way it does to its physical and biotic environment. Hence, the dilemma initially encountered in chapter 3, and reinforced in chapter 4, remains. Although there may be certain advantages to treating communities as nonlinear systems, these advantages may not provide the kind of theoretical advances that would be necessary to move community ecology toward the more rigorous, predictive science that some researchers would like.

Community Responses to Changing Environments

Although the conclusions reached in the last section, and those reached in chapters 3 and 4, may seem a bit gloomy, I think that rather than being cause for abandoning community ecology, they should be a cause for a redirection of its focus. I will spend a great deal of time discussing in later chapters one of the potential "new" foci for this redirection of attention, namely, macroecology. This does not mean, however, that the kinds of experimental and observational studies that have formed the bulk of community ecology in the past should be abandoned. In the balance of this chapter, I discuss approaches to community ecology on the local scale that might provide the basis for the kind of comparative community ecology first envisioned by MacArthur, and expanded upon by

21. Forecasting is used here in the sense of "predicting" future changes in densities of species within a community. This might be particularly useful for practical applications, where it is necessary to determine how a community will respond to changes in its environment.

Schoener (1986). Schoener's idea was that it might eventually be useful to classify different kinds of communities according to the types of mechanisms that determine their structure.

The basic idea that I consider in the rest of this chapter is that much can be learned about communities by examining how they react to changes in their environments. The idea of documenting changes in communities in response to changes in their environments has a long history in basic and applied ecology. Here I expand on this history by adding considerations from nonlinear dynamics and hierarchy theory.

Examining Nonlinear Responses

If communities are governed by systems that require complicated statistical techniques assuming nonlinear dynamics, then examining reactions of species to environmental perturbations, experimental or natural, using statistics that assume linear responses will provide limited information about the nature of how the community responds. Here I consider how we might go about examining nonlinear responses in communities.

A major difficulty encountered when analyzing nonlinear community responses is that considerably more data is needed than is conventionally collected. Usually, data requirements for studying nonlinear dynamics require time series with a minimum of thirty to forty data points, and this is probably far from optimal (see, e.g., Ellner and Turchin 1995). This is problematic because, traditionally, ecological systems have been studied for relatively short periods of time. In fact, some communities have dynamics that resolve themselves on time scales longer than the academic lifespan of the researchers studying them, so obtaining sufficient data to describe the nonlinearities in these communities is extremely difficult. Such considerations should clearly be part of study designs and funding decisions.

Despite these sample size restrictions, there is a sufficient number of ecologically significant time series available that it is important to examine the efficacy of statistical techniques that are not limited by assumptions of linearity for identifying components of community responses to change. I will not conduct here an extensive survey of the performance of techniques designed to identify chaos in ecological time series, primarily since such surveys have already been conducted (Ellner and Turchin 1995). Instead I will illustrate how such techniques can be used to provide interesting or new insights into a particular, well-studied system.

Consider the time series describing changes in density of *Dipodomys merriami,* Merriam's kangaroo rat, examined in the last chapter (fig. 4.1B). Recall that linear analysis of the community containing this species indicated that experimental effects of removing this species and its congeners could only be predicted for a limited period of time. The time series shown in figure 4.1B was the sum of densities on two control plots, where no removals had been done. During the first half of the time period shown, two congeneric species of *Dipodomys* coexisted with *merriami* on these plots. Shortly after the midpoint, however, the dominant, large-bodied competitor—the banner-tailed kangaroo rat, *D. spectabilis*—became extinct regionally (at about census 100). It is not immediately apparent from inspection of the time series in figure 4.1B that anything has changed. However, one would expect that if *D. spectabilis* were really a dominant competitor, its extinction would have a major impact on *D. merriami.*

To examine the possibility of some fundamental change in the dynamics of *D. merriami* when its dominant congener became extinct, I analyzed the first 102 censuses shown in figure 4.1B separately from the last 103. I used Ellner and Turchin's (1995) method to determine the largest Lyapunov exponent of each time series. The largest Lyapunov exponent is a statistic that measures the rate and direction that two initially similar trajectories from the time series might move over time. If the exponent is negative, the two trajectories will tend to converge, and any fluctuations in the time series would be due to extrinsic noise. If the exponent is positive, then the two trajectories are expected to diverge, implying that the time series is chaotic. Thus, this statistic is a way to differentiate chaos from local stability. The important point is that Ellner and Turchin's method allows there to be both an intrinsic, deterministic component and a random (possibly due to extrinsic factors) component of the system that drives changes in the species' density. Their procedure is to use a rather complicated goodness-of-fit criterion to separate the two components from one another, then to examine the deterministic component for the possibility of chaos.

The Ellner-Turchin method produced a rather interesting result. Before *D. spectabilis* became extinct, the time series for *D. merriami* densities was clearly chaotic, though mildly so (fig. 5.4). It is important to note that Ellner and Turchin (1995) chose criteria that were purposely conservative with respect to the possibility of detecting chaos. That is, their criterion, if biased, was biased against detecting chaos. So the result that *D. merriami* populations were chaotic before the extinction of a dominant competitor is a conservative one. Interestingly, for the second half of the time series, after *D. spectabilis* was extinct, *D. merriami* populations were strongly convergent. That is, the extinction of the banner-tailed kangaroo rat appeared to make the populations of *D. merriami* more stable. Much might be made of this result in light of competition theory. We currently do not have any good theoretical reasons to expect that extinction of

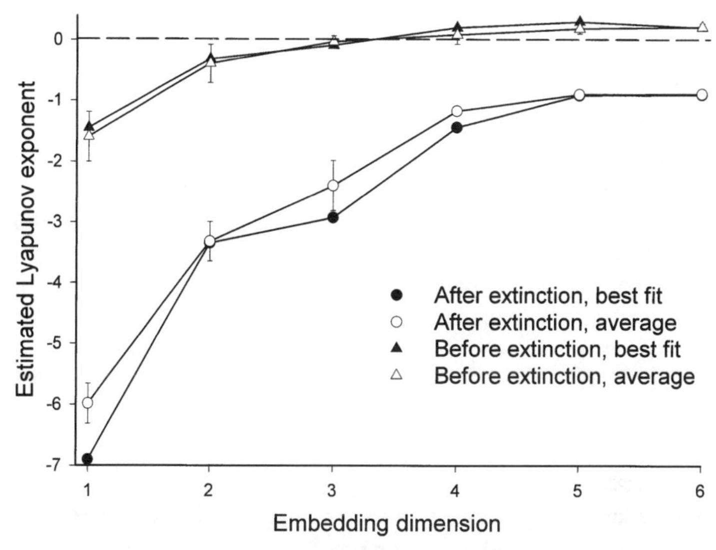

Fig. 5.4 Relationship between estimated Lyapunov exponents and embedding dimensions for two parts of a time series describing changes in the density of *Dipodomys merriami* in a Chihuahuan Desert ecosystem. During the first part of the time series, the dominant competitor *D. spectabilis* was present; during the last part of the time series, it was extinct. The embedding dimension represents the number of dimensions needed to capture the essential structure of the system. Estimates of Lyapunov exponents should level off when the appropriate embedding dimension is reached. Here, the embedding dimension is about four. The estimate of the Lyapunov exponent when the embedding dimension is reached is the best estimate of the global exponent of the time series. For the time series before extinction of *D. spectabilis,* the Lyapunov exponent is about 0.2, and for the one after extinction it decreases to about -1.3. Two methods of estimating the exponent are represented here. Closed symbols represent the best-fit model of the time series for each embedding dimension; open symbols represent the average of 20 best models; error bars represent one standard deviation.

competitors, particularly ones that have potentially strong effects on the species with which they interact, will result in stabilizing changes in the dynamics of the species that remain (although Holt 1992 suggested this possibility based on the qualitative argument that the product of two negative effects is a positive one, thus interspecific competition creates a positive feedback loop). However, the empirical result obtained here is intriguing, if nothing else.

Examining Multiscale Responses

If communities exist as hierarchies, then an explanation of patterns at one scale in the hierarchy must include reference to at least the next higher level and the next lower level in the hierarchy (Salthe 1985). However, levels at progressively distant scales from the level of interest (the focal level) require successively less reference in explanations of focal-level patterns. In practical terms, this means that attempts to describe community patterns must consider ecosystem-level and organismal-level processes, but may not require appeal to properties of such small entities as cells. Turner, Gardner, and O'Neill (1995) argued that such a formalization is too simplistic because entities such as communities cannot be considered to be contained within an ecosystem. Their concerns are essentially about the problem of closure I discussed above. Since there are no natural boundaries, treating a community as if it were contained within a larger system called an ecosystem potentially distorts or obscures our view of connections between community processes determining biodiversity with ecosystem processes determining rates of energy flows and nutrient cycling. However, their message still underscores the need for a more complex view of causality.

The expectation that processes at several scales are likely to be responsible for observed community patterns suggests that some protocol be established for making observations at different scales. Such a protocol need not require the use of new and unusual statistical techniques. In fact, it is desirable that any new protocol allow the use of statistical techniques already familiar to community ecologists. Such a protocol for analyzing communities has been developed by Allen and Starr (1982). The idea behind their method is that each type of calculation using raw data on species abundances in a community identifies a pattern that is affected by a set of processes operating primarily at a single scale. Thus, different types of statistical calculations can be viewed as "windows" that allow inferences to be drawn about processes operating at a limited range of scales. Allen, O'Neill, and Hoekstra (1987) called these different types of calculations *observation sets.* An observation set is a raw data set that has been subjected to a statistical transformation that eliminates the influence of certain processes on the raw data.

It is sensible to assume that different statistical calculations will emphasize the results of processes occurring on different scales because community kinetics are caused by many processes, each resulting from oscillations determined by the imprecise matching of energy used for maintenance by organisms in the community and loss of energy due to the second law of thermodynamics (Maurer 1987). Energy that enters the community is used for functions ranging from respiration and metabolism to resolution of interspecific interactions. The dynamics of each of these occur at different rates, so each has a different periodicity: some processes go through diurnal cycles, others through annual cycles, and so forth. Therefore, the raw data on changes in species abundances over time should contain many different periodicities, so that communities should exhibit what look like chaotic kinetics. Statistical calculations such as averages and variances, depending on how each observation is weighted, will tend to "ignore" or filter out some patterns that result from certain processes. The remaining patterns will show relatively little of the effects of the processes so filtered. Hence, each observation set resulting from a specific statistical technique will produce a pattern that is primarily influenced by processes going on at a limited range of scales. Ideally, one would hope processes occurring at only a single scale would show up in each observation set. However, both the natural heterogeneity of the community and the limitations of the statistical techniques available will in general ensure that there will not be a neat relationship between observation sets and scales. Realistically, we can expect that patterns in observation sets will be predominantly a reflection of a single scale or group of closely related scales.

The usual types of statistical calculations that are done on community data each weight the data so as to emphasize processes occurring at different scales, and therefore produce observation sets that may provide some information about how the community responds to processes occurring on different scales (Maurer 1985b). Some measures obtained from community data summarize the properties of the community as a single unit. Such measures include total biomass or total density, average biomass per species, and average biomass per individual in the community. I have referred to observation sets resulting from such measures as belonging to what might be called the community level of resolution (Maurer 1985b). Processes that affect each individual in the community in a similar manner should be evident at this level of resolution. At the other extreme, most methods of ordination emphasize the composition of the community by identifying species which are associated with one another across samples of the community (Ludwig and Reynolds 1988). These procedures emphasize the unique properties of individual species; hence, I have referred to observation sets generated from ordination methods as the individual species level of resolution (Maurer 1985b). An intermediate level of resolution in the analysis of community data is concerned with summaries of relative abundances of species, such as diversity indexes and rank-abundance curves. These analyses focus neither on the properties of the community as a whole, nor on the properties of individual species. Rather, they focus on the relative properties of species, regardless of the unique identities of individual species. Hence, this level of resolution is intermediate between the community and species levels.

Table 5.1 Observation sets used in an analysis of a grassland bird community in southeastern Arizona (see Maurer 1985b).

Observation Set	Level of Resolution	Criterion	Parameter or Data Transformation
1	Community	Density	Total density
2	Community	Density	Average density/species
3	Community	Biomass	Total biomass
4	Community	Biomass	Average biomass/species
5	Community	Biomass	Average biomass/individual
6	Intermediate		Species richness
7	Intermediate	Density	Number of common species
8	Intermediate	Density	Evenness
9	Intermediate	Biomass	Number of common large species
10	Intermediate	Biomass	Evenness
11	Individual species	Density	Ordination (principal component analysis)
12	Individual species	Biomass	Ordination (principal component analysis)

Using the organization of community measurements described above, I analyzed the dynamics of two bird communities during two breeding seasons. These bird communities existed in a highly variable desert grassland ecosystem, and there were within- and between-season changes in the ecosystem that had major impacts on the birds. To study these impacts, I developed 12 observation sets based on a number of data transformations (table 5.1). Because multiple processes operating at similar scales might influence the patterns at any given level of resolution, I calculated several observation sets for each level of resolution. I expected that each observation set at a given level of resolution would be affected by a different set of processes. In addition, I used different criteria to identify the importance of each species in the analyses. Since density is thought to reflect the population-level processes of productivity that are responsible for population dynamics, using density as a criterion for assigning importance to individual species in the calculations for some observation sets should reflect processes that have some affect on productivity. Biomass reflects something about the potential of the population to use energy (Peters 1983; Calder 1984), and thus I used it as a second criterion for assigning importance to individual species.

The results of the two-year study indicated that different components of the bird communities' responses to changes in the ecosystem could be identified at different scales (table 5.2). For example, at the community level of resolution, it was apparent that birds in mesquite-dominated

Table 5.2 Avian community properties in two different habitat types based on descriptions of the communities at three different levels of resolution. Numbers of observation sets from table 5.1 upon which the inferences are based are given in parentheses.

Level of Resolution	Habitat Type	
	Mesquite Savanna	Grassland
Community	1. Total density and biomass peak early in the season (1,3) 2. Average biomass of individual constant, smaller than in grassland (5)	1. Total density and biomass peak late in the season (1,3) 2. Average biomass of individual constant (5)
Intermediate	1. Increase in seasonal species richness in year of high rainfall (6) 2. Increase in number of large-bodied, rare species in year of high rainfall (6,9,10)	1. Bimodal seasonal pattern of species richness and evenness (6,7,8)
Individual species	1. Numerical domination of small-bodied insectivorous species early in season (11,12)	1. Numerical domination of large-bodied finches (11,12)

habitats reacted to seasonal changes in rainfall patterns by breeding early in the season, while birds in grassland habitats responded by breeding late in the season. At the individual species level of resolution, it was apparent that the species breeding early in the year in mesquite habitats were predominantly small-bodied insectivores, while those breeding later in the year in grassland habitats were predominantly larger-bodied emberizid finches. These changes in the bird communities were correlated with the different kinds of changes in productivity that occur in the different habitats (Maurer 1985b).

In summary, the practice of observing local communities using a hierarchical set of data transformations provides a basis for describing the relatively complex behavior that those communities are exhibiting. The fact that local communities are nonequilibrial, and therefore not structured in the traditional sense of MacArthur (1972) and others, does not mean that there are no biological interactions going on within those communities, nor does it mean that there are no methods available to describe their complexity. Using some of the techniques described in this chapter it should be possible to develop relatively standardized methods for describing local communities. This will then allow comparisons among different communities. It is in these comparative studies that I believe the general principles will begin to emerge.

Conclusions

At the scale commonly studied by ecologists, communities are collections of exceedingly complex associations of numerous species with each other and with their environment. The focus of community ecology in the past has been on the mechanisms of interaction among species in local communities. This focus has led to a number of successes, most importantly the empirical demonstration of the above definition of the interactions among species as exceedingly complex (see, e.g., Brown et al. 1986; Hairston 1989). I think that although such an emphasis should continue to occupy the efforts of some community ecologists, it is important to realize that there are limits to what can be discovered by relying solely on experiments. In fact, if the results obtained in this chapter for the desert rodents are any indication of the dimensionality of local communities, it may very well be that ecologists cannot measure enough variables to discover why the species they are studying exhibit relatively stable behavior (in a macroscopic sense). The rule that will emerge from continued study of local communities is that each is unique, despite the fact that we can easily enumerate the kinds of interactions among organisms that contribute to community dynamics. Thus, there will be limits on the generality of theories that we can produce to explain community structure and diversity. From a more practical viewpoint, there will also be limits on our ability to predict changes in communities, even if community dynamics are completely deterministic.

Because of the complexity of communities, it will be tempting to refer to their microscopic behavior as "random" or "stochastic." Although this may be operationally useful, I believe that it is conceptually inappropriate. If communities are governed by strange attractors, then attempts should be made to discover the macroscopic nature of the community by expanding the scale at which communities are viewed, rather than focusing on the apparent randomness and contingency they exhibit at local scales. Tools exist for studying long time series to determine if they have chaotic dynamics, but this information will probably be best used in a comparative sense. A population that has chaotic dynamics does not necessarily have a higher probability of extinction, so the presence of chaos should not be equated with instability within a community.

Although there is much to be learned by approaching communities as if they were complicated, nonlinear systems, many of the limitations inherent in the linear approach will remain. Such limitations arise because of the small scale at which communities are studied. Communities are not closed systems, so much of what is observed in a local community is determined by processes that transcend the arbitrary boundaries imposed upon it by the observations made on it. This limitation is inescapable. Ecologists simply cannot collect, much less conceptualize, the amount of information that would be necessary to identify precisely why species densities change the way they do. At the local scale, ecological complexity begets conceptual uncertainty, statistical variability, and theoretical unreliability. The kinds of regular structure that might serve as the basis for theoretical advances in ecology are not found at the scale of the local, two-to-three-year study. Hence, it is to the expansion of the spatial scale of community ecology that we turn now.

CHAPTER SIX

Macroecology: Expanding the Spatial Scale of Community Ecology

> As a result of this complex and endless interplay each species establishes temporarily an uneasy balance of numbers among all the others. The pattern of relative abundance of all the species in a mixed community (and all wild communities are mixed) is thus a synthesis of all the competition and cooperation, and all the difficulties and facilities, that have surrounded all the species of the community in the recent past. . . .
>
> C. B. Williams (1964)

One of my fundamental arguments up to this point has been that statistical regularities cannot be discovered in collections of objects that are of insufficient size, regardless of the complexity of their associations. Open systems that are of sufficient size may develop observable regularities, but these regularities can only be discovered if the system is viewed at the proper scale. A major problem for ecologists is determining exactly what scales are appropriate for documenting and understanding ecological phenomena (see, e.g., Lubchenco et al. 1991; Levin 1992; Turner, Gardner, and O'Neill 1995). Research in population and community ecology has focused on the level of the individual organism and its role in population and community processes. From this foundation, a rigorous experimental enterprise has arisen that, although it has the limitations discussed in previous chapters, is amassing a large database on mechanisms of interactions among species and how populations of species are regulated.

The growing database on plant and animal populations and communities has begun to provide views of patterns in assemblages of species that were not available earlier. This largely pattern-generating enterprise was dubbed "macroecology" by James Brown and I (Brown and Maurer 1989), and has led to the identification of a number of intriguing patterns (Brown 1995). Explaining these patterns is difficult because the standard tools of science for developing mechanistic explanations cannot be used. With few exceptions, there is no way to manipulate the biodiversity of a large geographic region. Another problem is that often the same pattern might be consistent with more than one process. Sorting among alternative explanations is therefore particularly hard.

The kinds of limitations discussed in the previous paragraph also apply to astronomy, but this has not prevented astronomers from learning a great deal about stars and galaxies.[22] Astronomy has a strong, quantitative theoretical foundation in physics, and this foundation allows fairly precise predictions to be made about what patterns can be expected in complex systems like galaxies (Smolin 1997). Is it possible that a similar relationship could emerge between macroecology and theoretical ecology? I think the possibilities are good that such a connection might be made. In this chapter I begin the quest to discover such a connection by examining some of the macroscopic patterns that need explaining and sketching how some of them might be explained.

Patterns in the Balance of Nature

Williams (1964) argued that statistical patterns in the abundances of groups of related species could be represented by a single statistical model: the log-series distribution. The title of Williams's book, *Patterns in the Balance of Nature,* was provocative: it implied that in all the diversity of nature there should be regular patterns and that these patterns resulted from the regulation of populations by their environments. The statistical patterns were striking, but the mechanisms were relatively unclear. This theme was reiterated by May (1986) in his address given on the receipt of the MacArthur Prize awarded by the Ecological Society of America. Patterns such as those that Williams and May described arise statistically in large collections of organisms. I shall examine several of these patterns and attempt to describe how their existence implies the operation of relatively general processes operating at large spatial scales.

The Spatial Distribution of Population Density

Population densities of most species within local habitats fluctuate strikingly both within a single year and from one year to the next. Habitats, in turn, are often distributed in patches across space. The correlation of population fluctuations in different habitat patches is often poorly understood. Some theoretical evidence suggests that populations that are patchily distributed in space but that are loosely connected by migration among patches might be more stable than a single population of large size, although this point has been hotly contested in the literature on conservation biology (Quinn and Hastings 1987, 1988; Gilpin 1988;

22. Brown (1995) attributed this point to a discussion he had with Robert MacArthur.

Wiens 1989a). The concern in this section is not to attempt to comment on this controversy, but rather to ask if there are any regularities that emerge when many populations are aggregated together in space.

Plant ecologists realized long ago that plants are not uniformly or randomly distributed across spatial gradients, but rather are distributed with densities highest at some point along the gradient and decreasing gradually in all directions away from that point (Whittaker 1967, 1975). This pattern has been documented for other groups of organisms (Hengeveld and Haeck 1982; Emlen et al. 1986) and is often seen in the geographic ranges of species (fig. 6.1; see also Hengeveld and Haeck 1981, 1982; Brown 1984; Hengeveld 1990; Curnutt, Pimm, and Maurer 1996). Gause (1930, 1932) conducted experiments to show that limitations imposed upon populations by an environmental gradient could result in a bell-shaped distribution across the gradient. Hence, both observational

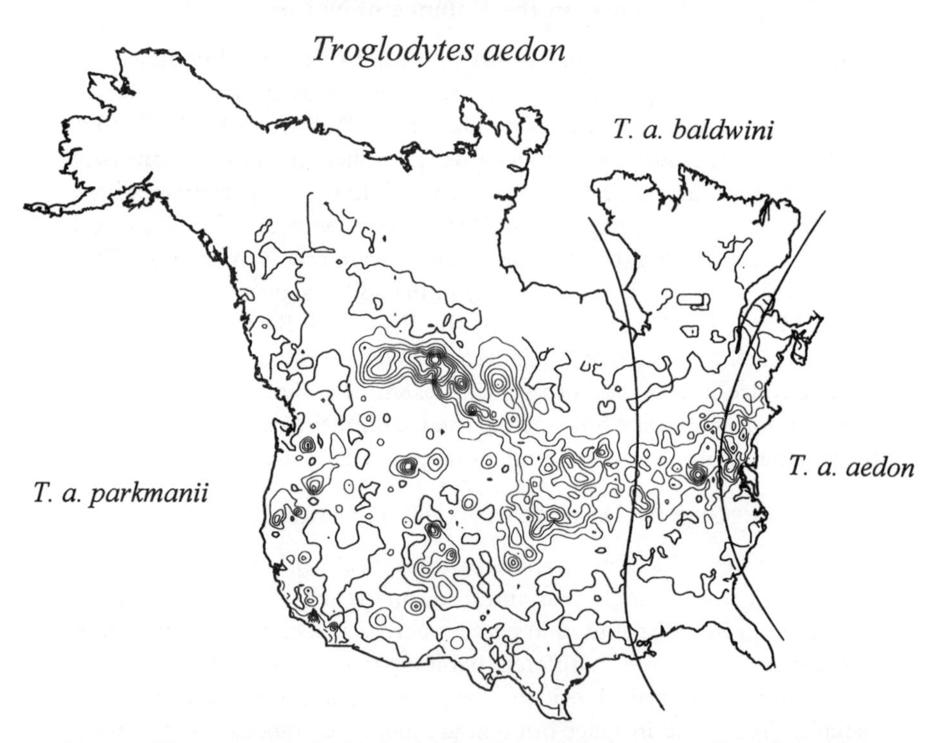

Fig. 6.1 The geographic distribution of abundance for house wrens (*Troglodytes aedon*) estimated from counts made by the U.S. and Canadian Fish and Wildlife Services. These counts, called the Breeding Bird Survey (BBS), have been conducted annually since 1965. Approximate boundaries between the three recognized North American subspecies (AOU 1957) are also included. Notice that there are several distinct regions of high abundance. Each of the subspecies has within its boundaries at least one of these regions.

and experimental data suggest that a bell-shaped spatial distribution of abundance is a general pattern repeated in a number of different organisms, regardless of their specific adaptations.

The important thing to recognize about this pattern is that it can be considered a statistical pattern across space. That is, any one population at any given point in time may depart from the pattern, but the overall pattern is a statistical one that holds when densities are averaged over time (Brown 1984). The explanation for this pattern requires that ecological conditions be spatially autocorrelated, and that there be some optimal set of conditions which limit the population processes responsible for population change (Hengeveld and Haeck 1981; Brown 1984; Maurer and Brown 1989; Hengeveld 1990). Brown and I (Maurer and Brown 1989) suggested that this spatial pattern is determined by a deterioration of ecological conditions away from the regions of high suitability. This deterioration of conditions can affect both the density-dependent and density-independent components of population regulation, so that the net effect is a lower average abundance in parts of the range with poor conditions. It should be noted that part of the environment experienced by a particular species is other species with which it interacts, so that biotic limiting factors can influence the distribution of a species. The pattern arises, then, through a statistical process of causality. Each individual organism contributes a small effect to the overall pattern based on its ability to contribute to population growth in its local habitat. The regularity emerges when sufficient numbers of individuals are found in a collection of populations in space.

Of course, not every geographic range shows the bell-shaped pattern. Large compilations of abundance maps of North American birds for winter (Root 1988c) and summer (Robbins, Bystrak, and Geissler 1986; Price, Droege, and Price 1995) show that a relatively large number of geographic ranges have more than one abundance peak. The common pattern for such multimodal abundance distributions is that of two or more bell-shaped distributions that abut one another (fig. 6.1). A number of factors could be responsible for multimodal distributions. For some species, each mode may represent a genetically distinct geographic form which is adapted to a different set of ecological conditions. For other species, preferred habitats may be patchily distributed geographically, so that different modes in the abundance distribution correspond to different regions where ecological conditions are optimal. Direct interactions of species in space may also modify the shape of the abundance distribution.

As an example of the complexities inherent in a geographical pattern of variation in abundance, consider the multimodal pattern for the house wren (*Troglodytes aedon*). House wrens are small insectivorous birds that require natural or human-made cavities in which to build their nests. Currently, there are three recognized subspecies in North America (fig. 6.1). The nominal subspecies occurs along the eastern coasts of the United States and Canada. A second subspecies (*T. a. baldwini*) is recognized that has an eastern border along the Appalachian Mountains and a western border formed by the Great Plains (AOU 1957). By far the most widespread of the three subspecies is *T. a. parkmanii.* This last subspecies extends across the Great Plains and up into the deciduous parklands of southern Canada. To the west, its range extends across Montana to the western coast of the United States. Southwest of the Great Plains, *parkmanii* extends through the Texas panhandle region, through New Mexico to southern Arizona. In southern Arizona and northern Mexico the taxonomy gets confused, and throughout Central and South America it is unclear whether there is a single large species, or many smaller species. This complex includes not only several species thought to be diagnosable, but also a number of localized forms and island forms. The three northern races described above are all migratory. They breed in the localities outlined in figure 6.1, but winter in the southern United States and northern Mexico (AOU 1983).

For the northern subspecies of *T. aedon,* we have a relatively large amount of data on geographical distribution of abundance. There are several concentrations of abundance throughout this range, and interestingly, each subspecies has at least one of these peaks within its boundaries. The boundaries of the subspecies' ranges correspond roughly to regions where house wren populations are relatively sparse. That is, between recognizable subpopulations of this species, there are regions where abundances are lower than regions well within each subpopulation's boundary.

The complexities of a geographical distribution of abundance, such as that for the house wren, do not refute the explanation that the pattern is the result of statistical regularities in the processes determining local population densities (Brown 1984). In fact, these deviations can be explained by relaxing the assumption of spatial autocorrelation of ecological conditions. The same demographic processes still operate to produce the multimodal pattern; the pattern results from including additional assumptions in the model. For example, a species may have a multimodal pattern because its optimal ecological conditions have a fragmented spatial pattern. The species itself may have more than one ecotype, so that optimal conditions differ among ecotypes. These additional assumptions extend the model to provide a more realistic description of geographic distributions of abundance without compromising the generality of its assumptions about the population processes that underlie these distributions. I will return to this pattern in the next chapter and discuss how we might begin to use the generalization to generate new predictions and hypotheses.

Patterns of Commonness and Rarity

In the previous sections it was shown that the distribution of abundance in a single species shows distinct regularities in form. Even species with fragmented geographical distributions have areas of relatively high abundance surrounded by areas of decreasing abundance. Are there general patterns in abundances that can be demonstrated in aggregates of populations of different, ecologically similar species? Hanski (1982a) was one of the first to show that there is indeed a striking pattern in assemblages of species distributed among patches across a landscape. Hanski was interested in examining the statistical distribution of the number of patches occupied by groups of closely related species. He reported several data sets that demonstrated that species which were most abundant in the patches or sites in which they occurred also occurred in the most sites. He went on to argue that because of this relationship between abundance and distribution, most species would tend to occur at nearly all sites or only rarely, but that there should be few intermediate species. Hanski called this hypothesis the core-satellite species hypothesis. Core species are those that are found on nearly every site, satellite species are those that are found on very few sites.

The core-satellite species hypothesis can be tested with appropriate data on the distribution of species across habitat patches or sampling sites. A number of methods have been devised to test the hypothesis. Hanski (1982b,c) demonstrated that some of the assumptions of his model were met for bumblebee and anthropochorous plant communities. Gotelli and Simberloff (1987) obtained data on the distribution of plant species among plots in the Konza Prairie Research Natural Area in Kansas. They found evidence that the distribution of species across the proportion of sites occupied tended to be bimodal within seven different soil types. They also showed that average density per plot occupied increased with increasing number of plots occupied. They concluded that this provided evidence of the core-satellite species hypothesis. Gaston and Lawton (1989) studied the distribution of herbivorous insects in stands of bracken (*Pteridium aquilinum*) in Britain. Although they found a positive relationship between average densities and number of sites occupied, they could find no evidence for bimodal distributions of number of species with proportion of sites occupied. Hence, they concluded that the core-satellite species hypothesis was not supported by their data.

They argued that this failure of the core-satellite model was not the result of the failure of their system to meet the assumptions of the model. Collins and Glenn (1990), on the other hand, showed that distributions of species numbers with number of sites occupied were bimodal both at the regional level and within local communities of grassland plants.

This lack of agreement among different data sets regarding the core-satellite species hypothesis is rather interesting. There may be some inherent difference between the sets of species in each of these studies. For example, herbivores inhabiting bracken may be governed by different sets of ecological processes than plants in tallgrass prairies. However, the positive relationship between distribution and abundance is tantalizingly general (see, e.g., Bock and Ricklefs 1983; Brown 1984; Bock 1984, 1987; Brown and Maurer 1987; Gaston 1988; Gaston and Lawton 1989; Owen and Gilbert 1989; as well as the studies cited in the previous paragraph). Perhaps the emphasis of Hanski's model on colonization and extinction processes is misleading. These processes are statistically determined by the activities of individual organisms undergoing their life histories. A consistent statistical trend of individuals at a given site failing to meet their life history requirements will ultimately lead to local extinction. Likewise, if this failure is consistent on a regional level, then such a species will have a higher likelihood of regional extinction.

Changing the focus from extinction dynamics to the ecological characteristics of individual species allows one to develop a more general explanation for both the conflicting results of tests of the core-satellite species hypothesis and the more general relationship between distribution and abundance. Brown (1984) suggested that the positive relationship between distribution and abundance could be explained by variation in the abilities of species to use available resources. According to Brown, species that use a wide variety of resources can find appropriate resources in more areas than species that use relatively few kinds of resources. This translates into both a higher likelihood of being found in samples of appropriate habitat and higher densities within those samples. Interestingly, in Maurer 1990a I showed that, based on these assumptions, it was possible to obtain both a core-satellite species pattern and a more uniform pattern in the distribution of species numbers with proportion of sites occupied by varying the level of dominance in resource acquisition of common species and changing the productivity of the habitat. In more productive habitats, an even distribution of species with proportion of sites occupied is expected, while in less productive habitats, where some species can monopolize resources, a core-satellite distribution is expected. This is one way to explain the discrepancy between the results of Gaston and Lawton (1989) for herbivorous insects and those of Gotelli and Simberloff (1987) and Collins and Glenn (1990) for plants in tallgrass prairies. If my analysis applies to these two groups of species, then bracken habitats are implied to be more productive with respect to the abilities of herbivores to use them than tallgrass prairies. Indeed, comparison of Gaston and Lawton's (1989) histograms with those from the studies of tallgrass prairie plants indicate that many of the bracken herbivore histograms appear to be relatively uniform.

The interpretation proposed in the preceding paragraph does not suggest that colonization and extinction processes are not important in determining the relationship between distribution and abundance. Rather, it suggests that these processes are the result of patterns of autecological success of individual organisms within different kinds of species. Hence, the positive relationship between distribution and abundance is a statistical regularity, evident in many different kinds of communities, that results from the many small events as individual organisms undergo their life histories. I also argued (Maurer 1990a) that the productivity of the environment constrains this pattern. I hypothesized that the steepness of the relationship between distribution and abundance should increase in more productive habitats. Hence, the pattern observed in any given community is determined both by the dynamics of small-scale processes such as resource use by individual organisms and by the constraining effects of larger-scale processes such as the productivity of a given ecosystem. Collins and Glenn (1990) interpreted their results similarly.

Hanski, Kouki, and Halkka (1993) attempted to test three different models of the relationship between distribution and abundance at the regional scale. In addition to the colonization and extinction approach and Brown's (1984) model of resource use, they examined a stochastic alternative. Wright (1991) showed that a simple Poisson sampling model could generate a relationship between average abundance and the proportion of sites occupied. Hanski, Kouki, and Halkka argued that the Poisson null model is inappropriate because populations are not randomly distributed in space (see also the previous section); they suggested instead a negative binomial sampling model. They found that in most cases the negative binomial model fit data sets on distribution and abundance better than the metapopulation or resource use models. This is an interesting result which suggests that the relationship between distribution and abundance may simply be a sampling artifact. Rare species may be found less often at local sites not because they are not there, but because they are harder to detect.

A major concern with Hanski, Kouki, and Halkka's (1993) conclusions is that it is not entirely clear that they can logically separate the three models they considered. They used a metapopulation model to derive predictions that is based on several unrealistic assumptions. For example, their model assumes that populations in all patches simultaneously achieve the same fraction of their respective carrying capacities (which is the measure of abundance that they use in the model). Populations must change this proportion in unison across the metapopulation. Of course, this assumption is unrealistic. If we reinterpret their model as depicting the *average* behavior of a population in a patch, then the model makes more sense. If, however, we assume that the model describes only the average behavior of the metapopulation, then we expect a great deal of stochastic variation in the outcome of these dynamics within an actual assemblage of species. Ergodic stochastic population processes, given enough time, converge to stable statistical distributions (Royama 1992). Hence, a flexible probability model like the negative binomial could easily describe the outcome of a stochastic version of Hanski, Kouki, and Halkka's metapopulation model. A similar argument could be made regarding the relationship between the resource use model and the negative binomial model. The resource use model is not as easy to separate from the other two models as Hanski, Kouki, and Halkka suggested. For example, a species that disperses readily (and hence results in a metapopulation that has relatively tightly connected patches) may be able to do so because it is capable of using either a widely available resource, or many different kinds of resources, or both.

Using a different approach, Holt et al. (1997) developed a simple population model that assumes species using the same set of habitats differ primarily in their density-independent response to environmental variation. In their model, species that have, on the average, higher density-independent growth rates will range over a larger fraction of habitats than those with lower density-independent growth. Their results add a more mechanistic interpretation to the qualitative explanations considered above.

A final point to note is that the spatial or temporal scale at which the relationship between distribution and abundance is examined may have profound implications for how we might need to explain the relationship (Gaston and Blackburn 1996a; Gotelli and Graves 1996). Early work by Williams (1964) suggested that patterns of commonness and rarity may be influenced by sampling effects related to the size of the area sampled. I return to this problem briefly in the next chapter. Perhaps distribution (e.g., geographic range size, per se) is not the best variable to use in examining the implications of patterns of commonness and rarity. There are some logical difficulties in determining exactly what the geographic range of a species (or its regional distribution) is supposed to be (Gaston 1994). In the next chapter I will develop an alternative concept of geographic range structure that might ultimately be more helpful in developing a theoretical understanding of the relationship between abundance and distribution, at least at geographic scales.

Density, Body Size, and Geographic Range Size

In the previous section it was seen that, within regional and local communities, ecological differences between common, widespread species and more narrowly distributed, rare species resulted in statistical regularities in the relationship between distribution and abundance. In this section, I will expand the scale of interest to ask whether there are patterns and statistical regularities that are evident in collections of species at larger scales, where ecological conditions change across some large geographic space.

Brown (1981) first documented a relationship between the size of a species' geographic range and its average body size in North American mammals (see also Brown and Gibson 1983). Subsequently, this pattern has been documented for North American (Brown and Maurer 1987) and Australian (Maurer, Ford, and Rapoport 1991) terrestrial birds. Gaston (1988) showed a strikingly similar pattern for British moths. Gaston and Blackburn (1996b) examined published data on body masses and range sizes for a number of different groups of animals. They classified studies into two broad categories: (1) comprehensive studies, that examined a geographic region that included within its boundaries most of the geographic ranges of the species being studied; and (2) partial studies, that included only part of the geographic ranges of species. They found that there tended to be significant positive relationships in comprehensive studies, but that a wide variety of relationships were obtained from partial studies.

Brown and I (Brown and Maurer 1987) went on to analyze the relationship between body mass and average population density for North American terrestrial birds. We found that, contrary to previous expectations (e.g., Damuth 1981), population density was not greatest for the smallest species, but instead reached its maximum for species of intermediate body size of around 100–200 g. Similar results have been obtained for several groups of insects (Gaston 1988; Gaston and Lawton 1989; Morse, Stork, and Lawton 1988). Damuth (1987) reported data from several sources and his data for herbivorous mammals from all continents indicated a pattern similar to the results obtained for North American birds and insects. However, his data set indicated that the minimum densities measured were not independent of body mass, but increased as species got smaller. It is likely that his data set was biased against obtaining reports of rare, small-bodied forms, as these are rarely studied in the ecological literature (Lawton 1989). When species assemblages have been extensively censused using standardized techniques, the resulting analyses usually agree that minimum density does not decrease with increasing body mass.

The relationship between average population density and geographic range size is an extension of the pattern seen among patches of similar habitat, but the amount of variation explained by a linear relationship on a log scale is small relative to the smaller-scale pattern (Brown 1995). Generally, there is a significant positive correlation between geographic range size and average density, but there are many more rare species with small geographic ranges than we would have suspected from a simple extrapolation of the distribution-abundance relationship at smaller spatial scales. The pattern for North American land birds indicates that most of the rare species with large body sizes are carnivores (Brown and Maurer 1987). However, the pattern is very similar within trophic groups, so it does not seem that the departure from the distribution-abundance relationship seen at the regional level is due to the inclusion of different trophic groups. Rather, it seems that this pattern arises from the inclusiveness of the data set: it represents collections of populations of different species across the entire continent, rather than within restricted geographic regions.

It might justifiably be asked whether or not these patterns would be expected from random combinations of these characteristics among species. To look at the properties of random combinations of these characteristics and to compare them to the actual observed combinations, I used data for North American terrestrial birds, which is the most complete data set that we currently have available. When an estimate of body mass for a species is multiplied by its density, one obtains an estimate of the average biomass per unit census area for that species. When the average density is multiplied by geographic range size, an estimate of the total number of individuals in the species is obtained. Finally, if body mass is multiplied by the total number of individuals, the total biomass of the species is estimated. The distributions for each of these variables for North American terrestrial birds are plotted in figure 6.2. It is possible to use the distributions of the derived variables for each species to test whether body mass, density, and geographic range size are combined randomly among species. To do this, I randomly arranged the values of each of these three variables for all species 500 times, so that the combinations of values were made without respect to species identities. With each permutation of values, I calculated estimates of average biomass per census unit, total numbers of individuals, and total biomass. This gave 500 distributions of average biomasses, total numbers of individu-

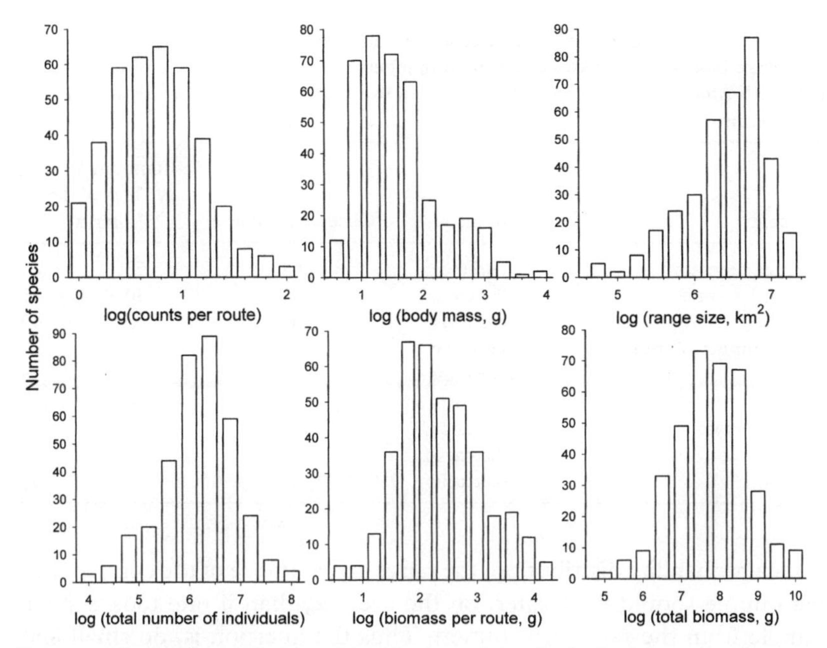

Fig. 6.2 Distributions of log-transformed values of density, body size, geographic range size, total number of individuals, average biomass per census route, and total biomasses among North American terrestrial birds.

als, and total biomasses. These distributions were then compared to the actual distributions of these variables to see if the observed distributions had any unusual properties. The observed distributions differed significantly from the randomly generated distributions (table 6.1). The observed distribution of average biomass per census route among species had a lower variance and was less positively skewed than was expected under the hypothesis of random combination of density and body size among species. Likewise, the variance of the observed distribution of total numbers of individuals among species was larger than that expected from a random combination of densities and geographical range sizes. The observed distribution of total biomasses was more negatively skewed than the random expectation.

More recently, some of these patterns have been examined using carefully constructed null models to separate "real" patterns from those that might be generated by a sampling effect. Blackburn, Harvey, and Pagel (1990) asked if it might not be possible that the relationship between abundance and body size can be accounted for by the fact that there are fewer very large and very small species than there are medium-sized species. If one randomly samples densities for a large number of species

Table 6.1 Comparison of randomly generated distributions with observed distributions of average biomass per census route, total numbers of individuals, and total biomass for North American terrestrial birds. All values were log-transformed before analyses were performed.

Variable	Statistic	Observed Value	Random Values (95% confidence interval)	
Average biomass	Variance	.51	.53	.63
	Skewness	.37	.40	.73
	Kurtosis	−.37	−.32	.49
Total number of individuals	Variance	.50	.40	.47
	Skewness	−.33	−.49	−.18
	Kurtosis	.31	−.16	.48
Total biomass	Variance	.89	.77	.92
	Skewness	−.10	−.01	.35
	Kurtosis	.03	−.27	.50

from a probability distribution, the range and variance of the corresponding sample should be greater, on the average, than if one took a small sample from the same distribution. Thus the question is, do small and large birds come from the same statistical distribution of abundances as medium-sized birds? Blackburn, Harvey, and Pagel developed several tests of this idea. Their most valid test was to randomly combine values of density and body mass among species (as I did above), and to use these to generate a bivariate "null" probability distribution of all possible combinations of abundances and body sizes. From this distribution, they showed that if one drew a medium-sized species at random (assuming that all species had equal densities, regardless of size), it would tend to have a higher density than either a large or small species would. This is important, because it shows that it is possible that even if there were no relationship between density and body size, it would still be possible to get a pattern similar to the one actually observed. However, we know that body size and density, at least for North American terrestrial birds, are not randomly combined among species (table 6.1), so Blackburn, Harvey, and Pagel's point is moot.

In a more detailed analysis, Blackburn, Lawton, and Pimm (1993) examined the upper boundary of the log abundance–log body mass plot. They considered the slope only for species with body masses larger than the body mass of the most abundant species. This ensured that the slope of the upper bound of log abundances would be decreasing with increasing body mass. They derived a number of values for the slope for several sampling models. They found that the slopes of the upper bounds of log abundance–log body mass plots for 22 of 24 data sets fell within the range of values from one or more of their sampling models. They concluded that there was no reason to invoke energetic explanations for the upper bounds when at least one of their sampling models could replicate the observed patterns.

All of the sampling processes envisioned by Blackburn, Lawton, and Pimm are based on two assumptions: (1) the statistical distribution of body masses is lognormal (1993, 697) and truncated at some small body mass; and (2) the mode of the body mass distribution corresponds to the body mass with the highest maximum log abundance. Neither of these assumptions is met for the North American terrestrial bird data. The distribution of body masses of this assemblage, and in fact of almost all groups of species, is not lognormal, but has a pronounced skew toward larger log body masses (Bonner 1988; Maurer, Brown, and Rusler 1992; Blackburn and Gaston 1994). Second, at least for North American birds, the modal body mass does not correspond to the body mass with maximum abundance (fig. 6.3). Despite this, the slopes of regressions of both the logarithm of number of species (using data above the modal body mass) and the maximum log abundance (using data above the body mass with maximum abundance) with log body mass are very close to −0.67 (fig. 6.4). Thus, the data for North American birds are not consistent with the sampling processes envisioned by Blackburn, Lawton, and Pimm. They are, however, consistent with some type of energetic limitation (Brown and Maurer 1987; Brown and Maurer 1989; Brown, Marquet, and Taper 1993), since logarithms of energy use and consumption by individual organisms vary with log body mass with a slope of approximately 0.67 (Brown and Maurer 1989; Peters 1983; Nagy 1987).

In a surprising and stimulating paper, Gaston, Blackburn, and Lawton (1993) showed that the relationships between engine size and abundance of different models (species) of automobiles in large samples taken at two parking areas in England were strikingly similar to those for body mass and abundance of species of animals. From an energetic standpoint, this result is not altogether unexpected. Gaston, Blackburn, and Lawton point out the similarities between cars and animals: both are open energy-processing systems subject to design constraints for efficiency and both experience differential survival of varieties (i.e., models of cars and species of animals). Since larger systems require more energy to construct (via factories or ontogenies), there are necessarily fewer of them.

Mechanistic explanations for species density distributions (Maurer, Brown, and Rusler 1992; Brown, Marquet, and Taper 1993) and relationships between body mass and abundance (Brown and Maurer 1986, 1987, 1989; Maurer and Brown 1988) incorporate known biological

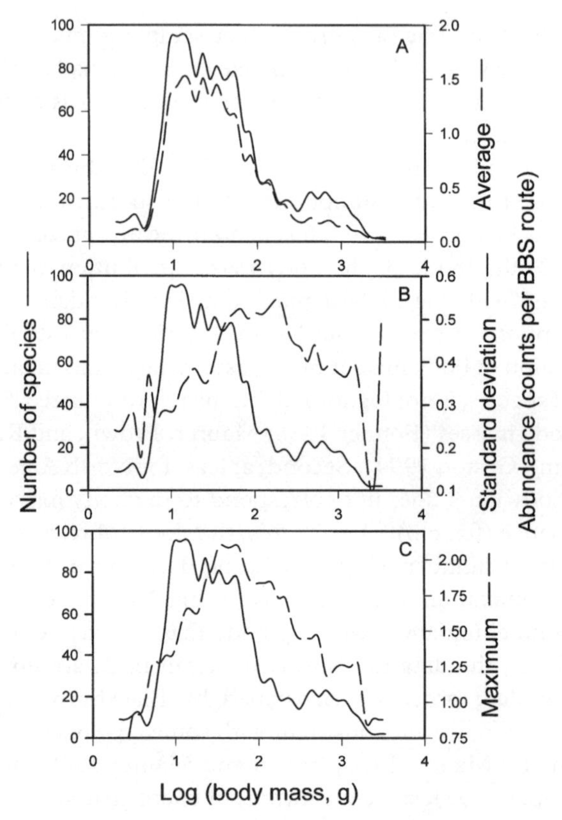

Fig. 6.3 Smoothed relationships between species density (*solid lines*) and characteristics of abundance variation (*dashed lines*) among species of North American terrestrial birds: *A,* average abundance; *B,* standard deviation of abundance; *C,* maximum abundance. Note that only average abundance peaks where species density is highest. Both the variability in abundance among species and maximum abundance peak at body masses greater than the modal body mass.

mechanisms to account for these patterns. For example, consider the relationship between geographic range size and body mass. The likelihood of extinction is influenced by total population size (Pimm 1991). Large species that have small geographic ranges must necessarily have small populations, and hence are more likely to become extinct than a small species with a geographic range of equivalent size (Brown and Maurer 1987). These mechanisms operating at relatively large spatial scales and long temporal scales are not commonly studied by ecologists and evolutionary biologists.

Although a great deal of effort has been expended to explain patterns of variation among species in body mass and abundance as the conse-

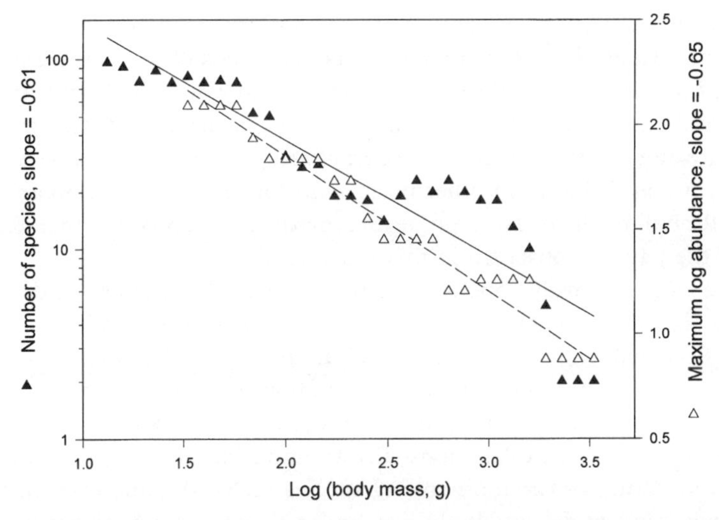

Fig. 6.4 Both log species number and maximum log abundance decline with a slope of approximately two-thirds for species of North American terrestrial birds. Points were obtained by taking the maximum species density and abundance for body masses greater than the body mass with the largest value for each variable. Thus, the modal body mass is smaller than the body mass with the highest maximum log abundance (see fig. 6.2 for complete distributions).

quence of passive sampling processes rather than biological mechanisms, many of the proposed sampling processes are based on assumptions that are violated in real data sets. Even though simulations of samples obtained from these processes have some similarities to observed data sets, the simulations are based on faulty assumptions, and hence it is logically inconsistent to accept qualitative congruence between simulated and observed patterns as evidence that the observed patterns occurred through the proposed sampling processes. This is a common fault of many null modeling exercises.

I should point out that the North American bird data set is unique in many respects, and much more likely to reflect mechanistic effects of nonrandom speciation and extinction and of design constraints. Abundances in that data set were obtained not at a few local sites, but across some three thousand census sites covering the entirety of the United States and Canada. Furthermore, the same census technique was used at each site. In contrast, most studies use either single censuses of entire communities at local sites (e.g., Brown and Maurer 1986; Morse, Stork, and Lawton 1988) or average abundances across a few local sites (e.g., Robinson and Redford 1986). The latter approach often compares abundance data collected using very different methods. Both of these approaches are much more likely to produce observations that can be attributed at least partially to sampling effects. Until data of comparable spatial extent to the Breeding Bird Survey (BBS; from which the North American bird data were obtained; Common Bird Censuses conducted in Britain are not sufficiently broad geographically to be comparable to the BBS) are available from other places and for other types of organisms, it will be difficult to test hypotheses regarding the processes that generated the patterns observed in that data set.

We can tentatively conclude, then, that the combinations of density, body size, and geographic range size observed in the North American terrestrial avifauna are not arbitrary, but are governed by biological processes that result in nonrandom combinations of these variables for species. How can we explain such nonrandom combinations?

Brown's and my explanations (Brown and Maurer 1987) for the nonrandom nature of variation among species of North American birds in combinations of density, body size, and geographic range size consisted of several related concepts. First, we hypothesized that constraints on the adaptations of organisms are imposed by their phylogenetic history; these design constraints prevented certain kinds of organisms from existing. Phylogenetic history imposes such constraints because, over the course of their evolution, species will accumulate certain adaptations that preclude others from evolving due to trade-offs of various sorts (Charnov 1993). The most obvious of these constraints is body size: a bird or mammal can only be so small without running into some serious metabolic and thermoregulatory problems. Some constraints, however, we thought to be more subtle. For example, because of the physiological mechanisms that underlie the digestive process, larger animals can eat food of lower energy concentration than smaller animals can. However, larger animals require more space. Hence, the maximum density achievable by species decreases beyond an intermediate body mass as species essentially "trade" biomass tied up in many small, energetically inefficient individuals for biomass tied up in fewer, more efficient individuals. Second, we hypothesized that some mechanisms that generated the geographic scale patterns that we documented were imposed on the collection of species by the physical characteristics of the continent. The most obvious of such physical constraints was continental size. However, Brown (1981) also indicated that the amount of energy available to organisms on a continent was an important physical constraint. Finally, we suggested that species with certain combinations of these properties made them more likely candidates for extinction. For example, species of large body mass can have small geographic ranges, but are likely to have large probabilities of becoming extinct. A recent example of such a species in North America is the California condor (*Gymnogyps californianus*), which experienced a precipitous drop in its total population size in recent years.

These explanations for the existence of such general patterns all incorporate in a specific way the general approach developed for statistical causality in chapter 2. In every case, the empirical patterns come from looking at large collections of organisms. Hence, organisms within each species contribute only a small effect to the overall pattern. Constraints on organisms within species are imposed by their phylogenetic history, by their participation in energy flow across geographic space, and by the nature of the physical environments in which they exist. The patterns are thought to result from consistent trends in the kinds of mechanisms in operation, so that over time certain processes dominate the behavior of these systems.

Body Size and the Allocation of Resources

Hutchinson and MacArthur (1959) noticed that the distribution of body masses among species in an assemblage of organisms that use similar resources tends to be a lognormal distribution. This means that there tend to be many more species of small body size in an assemblage than of large body size. They developed a model suggesting that the reason there are more species of small body mass is that such species can divide the habitat into more niches than larger species. Hutchinson and MacArthur's observations and model were nearly forgotten as ecology began to focus on small-scale, mechanistic explanations for community patterns. It was not until May (1978, 1986) gathered data on several different assemblages of species that interest was renewed in the pattern that Hutchinson and MacArthur had originally documented.

Since May's (1978) pioneering work it has become evident that the distribution of body masses in an assemblage of related species is nearly an empirical law (Bonner 1988). Bonner listed numerous studies that clearly demonstrated the approximately lognormal pattern. When Morse et al. (1985) showed that insects of different sizes could use the same habitat on different scales, it became apparent that something like Hutchinson and MacArthur's (1959) model of resource subdivision could explain the pattern. Damuth (1981) hypothesized on the basis of data on herbivorous mammals that the amount of energy used by populations of small-sized species was equal to that of larger-sized species because the smaller species tended to have higher population densities. Brown and I (Brown and Maurer 1986) cited evidence that within local communities this was generally not true. Damuth's original data set did not include complete censuses of local communities as Brown's and my data did. Harvey and Lawton (1986) proposed that, collectively, energy used by species of small body size might equal or exceed that used by large species because there are more small species. Brown and I (Maurer and Brown 1988) used data on the distribution of population energy use by North American terrestrial birds to show that neither Damuth's nor Harvey and Lawton's suggestions applied to that avifauna. Instead, they suggested that available energy was divided in a fractional manner among species of similar size. The size of the fraction taken by large species was larger, and hence the number of large-sized species that could be allocated shares of resources was smaller than the number of small-sized species. This conclusion, taken with Morse et al.'s data for insects, suggests that Hutchinson and MacArthur's original insight may have been essentially right. There are more small species because they can divide the resources available to them into smaller niches. The patterns of resource allocation documented in Maurer and Brown 1988 are not limited to terrestrial birds. Strayer (1986) showed that the distribution of population assimilation rates among species in a lacustrine zoobenthic community was qualitatively similar to the rates of population energy use of birds.

This talk of resource allocation might suggest to some ecologists the process of competition, where species are actively removing a share of the resources available to them and preventing other species from using them. Although this is certainly one mechanism that could explain the allocation of resources, it is not the only one. It may be that species of different body size allocate resources differently because individual species of small body size are physically constrained in their ability to dominate resources. Brown and I (Brown and Maurer 1989) argued that because species of small body size have higher metabolic rates, they require resources of higher energy concentration. Such species must specialize on patches of habitat with higher concentrations of energy. Hence, space is essentially more fragmented in terms of resource productivity for small species. Larger species can essentially average out spatial variation in resource quality. Furthermore, large species can usually move more effectively from one patch to the next. This means that populations of small species recognize the environment as more finely subdivided than populations of large species. Brown and I (Brown and Maurer 1989) showed that smaller species also tend to replace one another across space more often than large species. Brown and Nicoletto (1991) have reinforced this conclusion using a much larger database on communities of mammals in North America.

In a somewhat different vein, Dial and Marzluff (1988) suggested that large and small species differ in their ability to speciate and become extinct. They hypothesized that the observed lognormal distribution of body masses among species was a consequence of the balance between differential rates of speciation and extinction of small and large species. Brown, Rusler, and I (Maurer, Brown, and Rusler 1992) examined this idea in greater detail. We found that the distribution of body masses among species of mammals and birds on many continents was not lognormal, but was actually positively skewed on a logarithmic scale. Our explanation for this pattern invoked biases in the rates of speciation and extinction. Smaller species speciated faster than larger species and larger species became extinct faster, but these biases did not balance one another out, so instead of a lognormal distribution, a right-hand skew was produced over evolutionary time.

The common theme of explanations of patterns of resource allocation and body mass is the interaction of many organisms in large assemblages of species subjected to constraints placed upon them by body mass. These constraints regulate the distribution of niches that can evolve in an assemblage. The statistical perspective is extremely useful in explaining some intriguingly general patterns that exist in assemblages of species on geographic scales.

A Constraint Model for the Evolution of Body Size in Birds

Given the existence of the seemingly general patterns that I have discussed up to this point, the question naturally arises of how we might begin to explain these patterns using the theoretical framework discussed in chapter 2. The diversity of body masses within a taxonomic group follows a general pattern, as I pointed out in the previous section. If this pattern is a result of some set of processes operating within certain constraints on many different entities, then it is necessary to identify two things. First, we must be able to identify the underlying dynamical processes that are analogous to energy inputs (such as those which cause the motion of molecules of gas or the expansion of matter in the universe, as discussed in chap. 2). Second, we need to identify the constraints that operate on the behavior of the individual entities that make up the system (in this case, individual organisms within species).

In the model of body size evolution described in this section, the analogue of thermodynamic input into the system is Malthusian population growth. This is simply stating what Darwin and Wallace did in the last century, that is, the "drive" behind the evolution of body size is the unfaltering ability of organisms to replicate themselves. Constraints in the model come in two forms. In chapter 2 I pointed out that constraints can be both external and internal. Both of these kinds of constraint are included in the model. External constraints are those factors in the environment that selectively remove individuals from populations of species. This leads, of course, to natural selection, so up to this point I am not doing anything differently from previous evolutionary theorists. The real insight, however, is in how the internal constraints affecting the evolution of body size are defined. It has long been known that body size has important consequences for the population biology of a species (Peters 1983; Calder 1984). If this is really true, then natural selection may be constrained in different ways in species with different body sizes.

Why should we be bothered about this additional complexity of adding internal constraints to a perfectly good evolutionary model that is based on natural selection? The reason is that the addition of body size as a constraint to the model explains something natural selection cannot, namely, the skewed distribution of logarithms of body masses among species. I know it sounds somewhat heretical to suggest that there are some phenomena that result from evolutionary processes but that cannot be explained by natural selection. Of course, if the pattern among species can be generated by a random process, then we can disregard the pattern as epiphenomenal, and not worry about explanations that go beyond natural selection operating within species. In what follows I will show first that the skewed distribution of log body sizes among living species of birds is not due to a random evolutionary process, and thus is something that requires explanation. I will then show that the constraining effects of body size on the changes that can be produced by natural selection naturally lead to an explanation for the skewed log body mass distributions in birds and mammals.

Several explanations exist for the evolution of body size. Life history theory attempts to understand the evolution of body size at reproductive maturity by modeling the effects of life history trade-offs on fitness (Roff 1992; Stearns 1992). Although this approach successfully predicts adult body sizes within individual species, it has not yet been extended to show why the distribution of log body masses among species should be right skewed. Paleontologists studying the evolutionary history of body sizes in the fossil record have demonstrated that taxa usually originate from species that are relatively small and that over time both larger and smaller species arise in the lineage. These observations have generated a number of models for the evolutionary diversification of body sizes (see a discussion of some of these in Maurer, Brown, and Rusler 1992). The basic model assumes that diversification is random, and that physiological constraints determine the smallest and largest possible species within an evolving lineage. If the minimum body mass acts as a reflecting boundary, then a right-skewed log body mass distribution can be generated even if individual species are evolving independently and in different directions. This approach essentially ignores the mechanisms of body size evolution demonstrated by life history studies, although Brown, Rusler, and I demonstrated that a right-skewed distribution of log body masses is more likely to be produced if microevolutionary processes contribute to the diversification process. Neither life-history-based explanations nor random-diversification-with-constraints models fully account for the generality of right-skewed log body mass distributions.

Brown, Marquet, and Taper (1993) introduced the concept of reproductive power to provide a mechanistic approach to body size evolution that predicted the right-skew of log body mass distributions. They defined reproductive power as "the rate of conversion of energy into useful work for reproduction" (575), and considered it to be an alternative definition of fitness. The quantitative definition for reproductive power was determined from a steady state model incorporating two physiological processes: (1) the ability to turn resources into biomass (acquisition, A), and (2) the ability to turn accumulated biomass into offspring (conversion, C). At steady state, reproductive power, P, expressed in units of energy per individual per unit time, is

$$P = \frac{AC}{A + C}. \quad \textbf{(6.1)}$$

Allometric expressions for acquisition and conversion are available in the literature for mammals, so reproductive power can be expressed as a function of body mass. The shape of the reproductive power function plotted against log body mass is right-skewed, and has a maximum that corresponds to the modal body size for North American mammals.

Logical difficulties arise when we equate reproductive power with fitness. Average fitness of a population is necessarily maximized by the process of natural selection. Reproductive power has no similar relationship to a process that requires its maximization. Often, the component processes of reproductive power are involved in trade-offs, so that as one increases, the other decreases; but there is no indication that these trade-offs must maximize reproductive power. Even if reproductive power can be shown to be maximized by natural selection, or by some set of physiological processes related to fitness, most species of North American mammals do not have body masses that optimize reproductive power. It is possible to envision an evolutionary process whereby competition might prevent species from reaching maximum reproductive power, but this would require competition between groups as divergent ecologically as carnivores and rodents. If reproductive power cannot be considered to be an alternative formulation of fitness, it must still be related to fitness. Fitness is the rate at which a phenotype increases or decreases in a population. The physiological efficiency with which a phenotype obtains resources and transforms them into offspring clearly will influence its ability to increase in a population. However, it is important to separate physiological processes from their population genetic consequences, and hence reproductive power and fitness must describe different aspects of the mechanisms that lead to evolutionary change.

How, then, can the relationship between reproductive power and the number of species of a given body size shown by Brown, Marquet, and Taper (1993) be explained? It is possible that this relationship is simply fortuitous, and that evolution is random with respect to body size. To examine this possibility, I obtained body mass data for 6,217 bird species (Dunning 1993). There are about 9,672 recognized species, so this sample represents 64% of all known living species. The distribution of log body masses of species of birds is right-skewed, and it is unlikely that this skew is due to the fact that not all species of birds are included in the sample (Blackburn and Gaston 1994).

If the direction of body mass evolution within a lineage is random over evolutionary time, then the distribution of body masses among species within the lineage taken at any given point of time should be lognormal (Hutchinson and MacArthur 1959; Stanley 1973). If the smallest possible body mass achievable by the lineage acts as a reflecting boundary, then a random diversification process will generate a distribution of body masses on log scale that is right-skewed (Stanley 1973; Gould 1988). The distribution of body masses of birds is clearly not lognormal (Bonner 1988; Maurer, Brown, and Rusler 1992), so it is unlikely that diversification of body sizes in birds was completely random. The second alternative can be tested using the subclade test (McShea 1994). If diversification of body size is random, but coupled with a reflecting boundary at small body sizes, then monophyletic taxa that are relatively distant from the lower body mass boundary should have distributions that are symmetric (McShea 1994). I used a recent classification scheme based on DNA hybridization (Sibley, Ahlquist, and Monroe 1988) to classify each species in the sample into one of 23 orders. Thirteen of these orders had average abundances greater than the average for all species. The average skewness for these orders was significantly different from zero (table 6.2). Hence, the hypothesis that the skewness of the log body mass distribution for the world's birds was due to random diversification of body sizes in birds coupled with a lower reflecting boundary is rejected.

The body mass distribution of living birds was not generated by a random diversification process. This should be reflected in the history of body mass changes in birds. Since the fossil record of birds is not yet

Table 6.2 Results of the subclade test for skewness of body mass distributions in 13 orders of birds.

	Point Estimate*	95% Bootstrap Confidence Interval*	
		Lower	Upper
Average skewness	.36	.23	.50
Difference between average skewness and skewness for all species (0.79)	−.43	−.43	−.50

* Since there is an unknown correlation among orders due to phylogenetic relationships, conventional statistical tests cannot be used (Harvey and Pagel 1991). Because the sampling distribution of skewness estimates is unknown due to this correlation, it was estimated using the bootstrap technique. This technique generates a sampling distribution from the empirical distribution function. Average skewness was the average of two thousand bootstrap samples; confidence intervals were obtained by taking appropriate percentiles from two thousand bootstrap samples (Efron and Tibshirani 1993).

detailed enough to test this directly, I obtained an estimate of the history of body mass evolution in birds using an estimate of the phylogenetic relationships among 23 orders (Sibley, Ahlquist, and Monroe 1988). I created a two-state character for each of the 23 orders of birds that indicated whether the average body mass of that order was larger or smaller than the average bird. The phylogenetic reconstruction, assuming a minimum number of character state changes, indicated that taxa with large average body sizes were always derived from taxa with small body sizes (fig. 6.5). This implies a net directionality to the diversification of body sizes of birds from small to large. Note that this is only a net direction, since within orders there are species that are smaller than the average bird for that order.

I conclude that the relationship between species number and reproductive power is not fortuitous nor due to a random diversification process. The relationship between species richness and body size is due to a process that results in a net directionality toward larger body size. This relationship is also concordant with the relationship between reproductive power and body size. I estimated the relationship between reproductive power and species richness in birds by obtaining allometric estimates for acquisition (Ricklefs 1974) and conversion (Farlow 1976) for birds (Maurer 1998a). The relationship between reproductive power and species richness is different for large and small birds (fig. 6.6). Species richness declines more rapidly with declining reproductive power for small birds than for large birds.

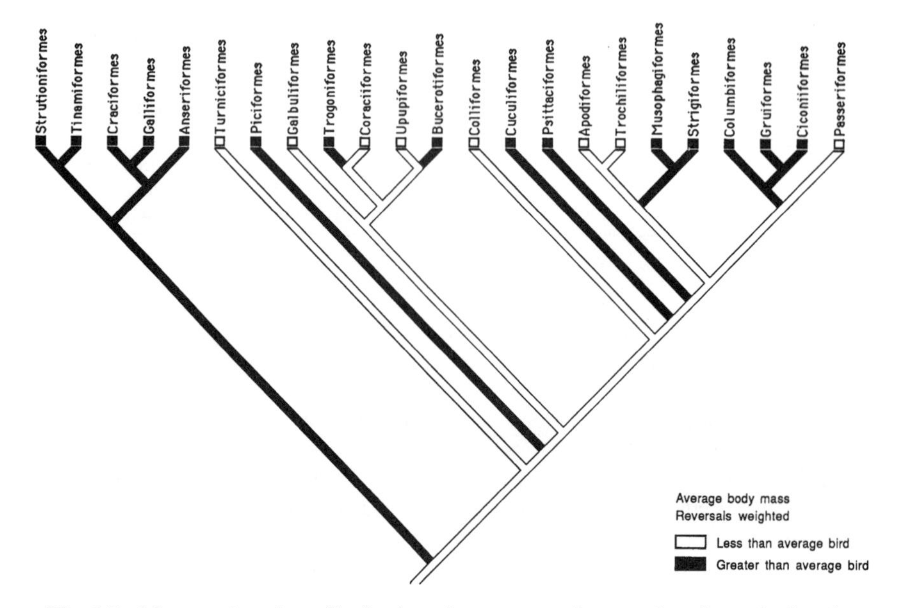

Fig. 6.5 Most parsimonious distribution of a two-state character based on whether the average body size of an order was greater or smaller than the average for all birds. *Filled lines* indicate lineages estimated to have had average body masses greater than the average for modern birds; *open lines* indicate lineages estimated to have had average body masses less than the average for modern birds. Relationships among 23 orders are from a phylogeny based on DNA hybridization. Parsimony analysis was done using MacClade (Maddison and Maddison 1992). Transitions from large to small taxa were weighted as having two steps since such transitions in the fossil record are relatively rare (only one such transition occurred in the reconstruction, with the fewest number of steps obtained when equal weights were assigned to transitions). Note that this reconstruction has no instances in which a small taxon originated from a large taxon.

To investigate why the relationship between diversity and reproductive power is different for small and large birds, I reexamined the definition of reproductive power given by Brown, Marquet, and Taper (1993). Recall that reproductive power is determined by two processes: acquisition of biomass and conversion of this biomass to offspring. The effect of each of these processes on reproductive power was examined by plotting the partial derivative of reproductive power with respect to each component process (Maurer 1998b). For both birds and mammals, changes in reproductive power of small species are dominated by the acquisition component (fig. 6.7). This means that changes in the ability of a small species to convert biomass into offspring will effect little change in reproductive power. Likewise, changes in reproductive power for large species are dominated by the conversion component (fig. 6.7). Thus, changes in the ability of large species to acquire and transform resources into

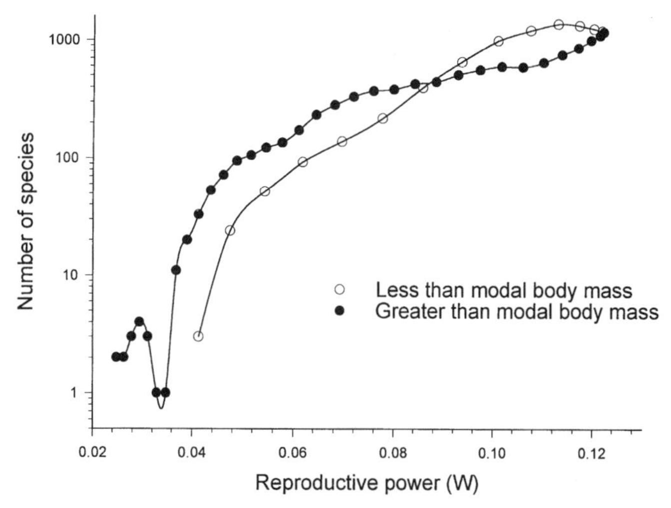

Fig. 6.6 Relationship between reproductive power and species richness for the world's birds. The relationship was constructed using the following smoothing algorithm. I chose an interval of 0.318 on a log scale (this represents a constant ratio of about 2 between the upper and lower bounds of the interval on an arithmetic scale). I then ranked species from the smallest to largest log body mass. Setting the lower bound of the interval on the smallest species, I counted the number of species within the interval, and calculated the reproductive power for the midpoint body mass of the interval. I then increased the lower and upper bounds of the interval by a small amount and repeated the calculations. This procedure was repeated 50 times, so I had 50 pairs of values for number of species and reproductive power.

biomass will have little impact on reproductive power. Reproductive power is constrained by different processes in small and large species. Note that species of intermediate size are relatively unconstrained by either component of reproductive power.

Reproductive power thus seems more likely to be involved in constraining the evolutionary process than in directly determining its outcome, as does fitness. Increasing the ability of an individual to process resources and convert them to offspring will increase the fitness of that individual, so phenotypic changes that increase fitness must also in some way increase reproductive power. But changes in reproductive power are constrained by body size. Heritable phenotypic traits that increase the ability of a small organism to transform its accumulated biomass into offspring will have little effect on reproductive power, and will therefore have little effect on fitness. In order to increase fitness in small organisms, phenotypic changes must increase the ability of the organism to acquire resources and transform them into reproductive tissues. In a like manner, phenotypic changes in large organisms must increase the ability of the

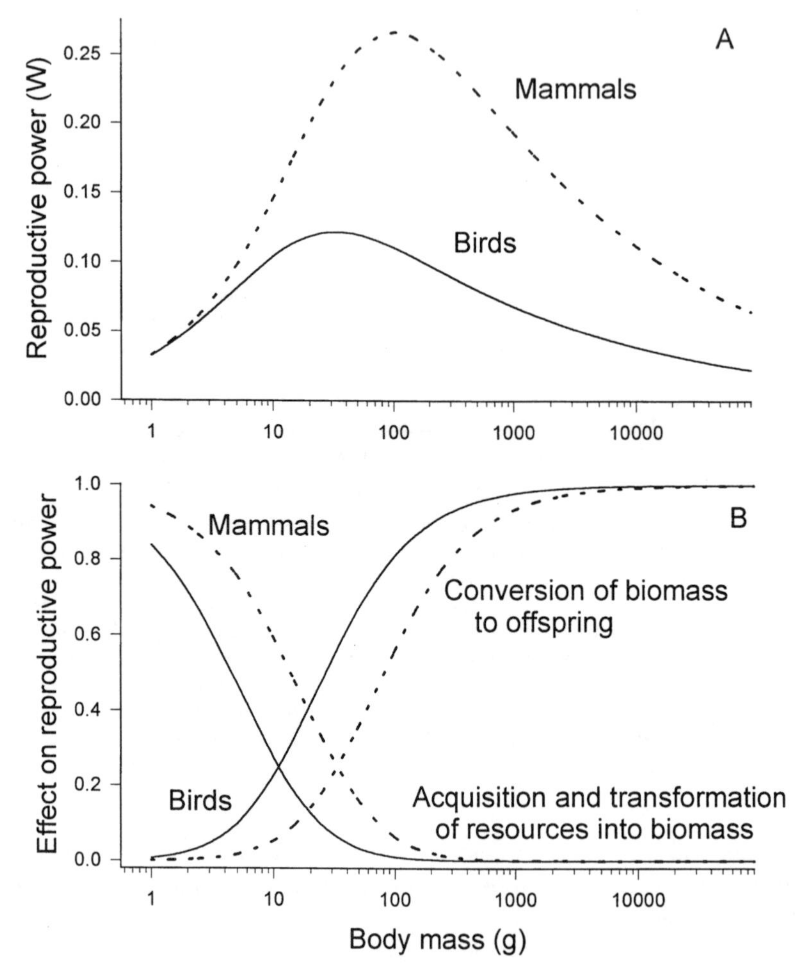

Fig. 6.7 *A,* distribution of reproductive power with body mass for birds (*solid line*) and mammals (*dashed line*). *B,* estimated partial derivative of reproductive power with respect to acquisition and conversion for birds (*solid lines*) and mammals (*dashed lines*).

organism to transform its accumulated biomass into offspring if those changes are to increase fitness.

The distribution of reproductive power for birds differs in two ways from that for mammals (fig. 6.7). First, the body mass at which reproductive power is maximized is less for birds (33 g) than for mammals (100 g). Second, except for the very smallest birds and mammals, a mammal of a given body mass will have higher reproductive power than a bird of similar size.

Mammals are generally able to achieve higher population densities than birds (Brown 1995). The exception to this is the smallest birds and mammals, which converge on the same maximum densities (Brown 1995). Both the generalized pattern and the exception for the smallest species may be direct consequences of the distributions of reproductive power among species of birds and mammals. Given similar resources, mammals larger than the smallest species should be more efficient at turning those resources into offspring than birds. This ability should translate into higher population densities, all else being equal.[23] However, for both the smallest birds and mammals, reproductive power is nearly identical (fig. 6.7), which implies that small mammals have no greater ability to turn resources into offspring than small birds, and consequently population densities of small mammals and birds are nearly identical.

Although mammals generally have higher population densities, there are more than twice as many bird species as mammals species (Sibley and Monroe 1990; Nowak 1991). Thus, the higher reproductive power of mammals does not lead to higher mammalian diversity. Birds and mammals probably have very different likelihoods of being fossilized, so it is not clear whether these diversity differences are typical. If they are, this suggests the interesting possibility that the relatively low population densities that birds are able to maintain might be related to their rate of diversification. If bird and mammal species form primarily by geographic speciation, then the lower population densities of birds might predispose their populations to become more isolated. Although such isolation might be partially offset by the ability of individual birds to travel across barriers, birds may be less capable of maintaining viable populations in the harsh regions that separate populations, enhancing the likelihood of bird populations' becoming sufficiently isolated to stimulate speciation.

The differences in reproductive power between birds and mammals are consistent with Brown's (1995) suggestion that the costs of flight have profound consequences for birds. For a bird to fly efficiently, it must allocate less of its structure to internal organs than a mammal, thus limiting the kinds of resources it uses to those that require less internal processing. Since birds use fewer resources, they maintain lower population densities than mammals (Brown 1995). This is consistent with the results on reproductive power obtained in this study.

Because of the constraining effects of body size on fitness, species that are very large or very small are rare because they are subject to strict constraints. They generate lower absolute amounts of reproductive

23. That is, if density dependence, mortality, and resource availability are not tremendously different between birds and mammals. If this is not true, then this explanation may need to be modified to account for those differences.

power (fig. 6.7), and any phenotypic changes are constrained by one or the other component of reproductive power. Species of intermediate body size are more common because they suffer relatively few constraints and can generate much greater amounts of reproductive power. Since reproductive power is less constraining for large species, there is a statistical tendency for diversification to proceed toward larger organisms, consistent with the proposed evolutionary history of body size changes for birds (fig. 6.5).

The role of reproductive power as a constraint on diversification applies not only to microevolutionary processes, but to macroevolutionary processes as well. Reproductive power can affect the outcome of speciation. A successful speciation event requires not only genetic divergence, but also the founding of a viable population (Eldredge and Gould 1972). Although genetic divergence might occur without the influence of reproductive power, the establishment of a viable population very clearly will be influenced by reproductive power. Furthermore, since extinction is basically a demographic process, consistent failure to accumulate enough resources to produce sufficient offspring will influence the extinction process. Consequently, both anagenetic (natural selection) and cladogenetic (speciation and extinction) processes are affected by reproductive power.

In summary, there is strong evidence that the evolution of body size in birds does not proceed haphazardly. It is a directed process, but the direction does not come from a simple adaptive goal, such as evolution toward an optimal body size. Rather, the processes that determine the tempo and mode of evolution, namely natural selection, speciation, and extinction, are constrained in their outcome by the body masses of the particular set of species involved. Because the direction and rate of evolution are constrained by different factors in species of different sizes, the net effect is that body size evolution must occur within a set of constraints that lead to the right-skewed body mass distribution characterized for so many different kinds of organisms. Thus, a general ecological pattern may ultimately be explainable by a relative simple set of general principles.

Conclusions

We have seen in this chapter that intriguingly general patterns exist in large collections of species, at least for the few cases in which they have been examined empirically. The explanations that have been put forward for many of these patterns incorporate elements of the general model of statistical causality developed in chapter 2. These patterns and their explanations provide a different perspective on the nature of biological diversity than has been obtained from the more reductionist approaches developed from the study of local communities (Brown and Maurer 1989).

The existence of regular patterns in relatively large collections of organisms suggests that a fruitful avenue for future theoretical development is one that explicitly incorporates a statistical perspective regarding the explanation of patterns in communities. This perspective can extend beyond the spatial and temporal domains of traditional community ecology to give us insights into the relationship between diversity and productivity in a patchy landscape and into the evolutionary diversification of clades of species over time.

CHAPTER SEVEN

Geographic Range Structure: Niches Written in Space

> There is something unsatisfactory in such abstractions that seem rather far remote from the conditions actually met in nature. But it must be remembered that such abstractions are a necessary, and, as experience has abundantly shown, a very effective aid to our limited mental powers, which are incompetent to deal directly with unexpurgated nature in all its complexity.
>
> A. J. Lotka (1925)

In this chapter, I return to a pattern introduced in the previous one, namely, the distribution of abundance of a species in space. This pattern is one that is well documented empirically, and one about which we have at least some theoretical understanding. I will illustrate how this pattern can be viewed from the statistical perspective that I have been advocating throughout. It turns out that approaching the problem from the statistical viewpoint provides new opportunities to understand the mechanistic basis of the pattern. It leads directly to an expansion of current theory regarding spatial population dynamics to include geographic ranges. If the approach outlined in this chapter is eventually used to examine data from species other than birds, which have the best geographic data on population abundance currently available (Maurer 1994), then the generality of the patterns discussed in this chapter can be assessed. More important, however, are the implications of the descriptions of geographic ranges presented in this chapter for how we view ecological mechanisms that contribute to evolutionary processes, such as speciation and extinction. I examine these implications in chapter 9.

Toward a Statistical Theory of Geographic Range Dynamics

It would be most desirable to be able to develop, from first principles, a theory that describes why population density is distributed the way it is across a species' geographic range. I am not sure, however, that we have enough data or the right combination of theoretical tools to provide such a theory. Interestingly, attempts to develop such a theory have a long history in ecology. Initially, the interest of many ecologists was to explain why population densities change across space. Grinnell (1917) used the term "niche" to refer to the ability of species to be found in only certain places across space (James et al. 1984). This gave rise to attempts by researchers such as Gause (1930, 1931, 1932) and others to show that species tend to be unimodally distributed along axes that represented important ecological conditions. The awareness that many factors influenced the population dynamics of competing species led Hutchinson (1958) to conceptualize the niche as a region in a multidimensional resource space within which a species was able to maintain populations. He used this concept of the niche to describe how competitive interactions might reduce the range of resources that a species would use within this space. In concentrating on resource space, less emphasis was placed on the particular location where a species was able to maintain populations.

Although studies of the niche have played an important role in the development of community ecology, and are likely to do so in the future (Leibold 1995), the concept of a niche has become abstracted from its original, spatially explicit context. It is incumbent upon the field of ecology to return to the spatially explicit formulation of the niche concept now that it is becoming ever clearer that ecological phenomena are generally spatially dependent. Some have pointed out that with the advent of modern community ecology, the role of history in ecology was "eclipsed" by an atemporal, mechanistic view of ecological theory during the 1960s and 1970s (Kingsland 1995; Brooks and McLennan 1991). It is probably also valid to suggest that this mechanistic approach eclipsed the role of space as well. That is not to say that a mechanistic view of ecology is necessarily bad. Rather, it is important to begin to examine mechanisms within a spatial context.

There are numerous ways to reinstate space into ecological theory (Kareiva 1994). Most of these are focused at the micro- or mesoscale. In what follows, I examine one way to make macroecology spatially explicit. I begin by returning once again to A. J. Lotka's ideas. This time, I consider a part of his book that has rarely been discussed since it was published. His ideas provide a springboard for examining the nature of geographic populations.

Lotka's Intensity Law for Biological Populations

Lotka's vision of biological evolution extended far beyond the description of population dynamics I discussed in chapter 4. In his far-ranging discussion he anticipated ideas as diverse as life history optimization, optimal foraging, ecosystem dynamics, and hierarchical organization. These ideas were all developed from a thermodynamic perspective, that is, by considering the nature of energy flows through biological systems.

The quote at the beginning of this chapter was taken from Lotka's chapter 23 (1925, 300–321). In that chapter, he points out that the equations for "interspecies evolution" that I described in chapter 4 (eq. [4.1]) contain certain parameters that reflect the effects of the environment on the rate of population change. If we let $P_1, P_2, \ldots, P_n$ represent these parameters, then equation (4.1) becomes

$$\frac{dN_i}{dt} = N_i f_i\,(N_1, N_2, \ldots, N_s, P_1, P_2, \ldots, P_n). \tag{7.1}$$

My treatment of equation (4.1) assumed that these environmental parameters were constants, and so could be ignored. But, as Lotka so forcefully pointed out, at some point the complexities inherent in the relationship between population dynamics and the environment must be considered if we are to develop a more complete theoretical framework for understanding population dynamics in space and time. Lotka made some very general observations about how this might work and, in doing so, elaborated a concept he called an "intensity law." In what follows, I use this law as the basis for a discussion of geographic ranges.

Lotka observed that biological populations occupy geographic space in much the same way as an ideal gas might occupy a particular volume. The size of the volume occupied by such a gas depends on the relationship between the energy content of the gas (as evidenced by the motion of its constituent molecules) and the constraining force of the container the gas occupies. If p_E is the external pressure exerted upon the gas by the container and p_I is the internal pressure generated by the energy content of the gas, then the rate of expansion (or contraction) of volume occupied by the gas has the same sign as the difference between the internal and external pressures, or

$$\frac{dv}{dt}\begin{cases} > 0 & \text{if } p_I - p_E > 0, \\ = 0 & \text{if } p_I - p_E = 0, \\ < 0 & \text{if } p_I - p_E < 0, \end{cases}$$

where v is the volume occupied by the gas. Likewise, the area occupied by the population of a particular species is determined by a "population pressure," ρ. The size of the area occupied by a population, a, increases when the internal pressure, ρ_I (generated by an excess of births over deaths), forces more individuals to emigrate to surrounding geographic regions. The environment exerts an "external pressure," ρ_E, by destroying individuals faster than they are produced. The area occupied by the population thus changes as

$$\frac{da}{dt}\begin{cases} > 0 & \text{if } p_I - p_E > 0, \\ = 0 & \text{if } p_I - p_E = 0, \\ < 0 & \text{if } p_I - p_E < 0. \end{cases} \quad \textbf{(7.2)}$$

As with the ideal gas law, Lotka argued that the existence of the conditions implied in equation (7.2) suggests a reasonable relationship between the dynamic consequences of life history processes and the space occupied by the population. In fact, he suggested that the product of population pressure (measured in units of individuals per unit area per unit time) and area was equivalent to the rate of population change, or

$$a_i \, \rho_i = N_i f_i. \quad \textbf{(7.3A)}$$

Rearranging gives

$$\rho_i = \frac{N_i}{a_i} f_i, \quad \textbf{(7.3B)}$$

where N_i is the number of individuals in the population of species i, and f_i is the per capita rate of change of that species (the same function f given in eq. [7.1]). The relation in equation (7.3B) is isomorphic to the ideal gas law, where ρ_i is analogous to pressure and f_i is analogous to temperature. In other words, the "energy" of the population is determined by the rate of production of offspring, and the greater this energy, the greater the power the population exerts to expand its geographic range.

Lotka's isomorphism between ideal gases and populations goes beyond the outward resemblance of equation (7.3B) to the ideal gas law. Lurking beneath this rather calm exterior is the implication that population dynamics are like statistical mechanics; that is, the myriad of activities that each individual organism in a population goes through during its life has a small effect on population dynamics in space and time, but the sum total of all these activities across all individuals in the population within a discrete geographical region can be measured by macroscopic variables. In Lotka's formulation, these macroscopic variables are population pressure (which might be interpreted as the average rate of dispersal per unit area) and the per capita rate of population change. The per capita rate of change of a population has been the object of intensive study in ecology for many years, but population pressure has rarely been measured. There are many estimates of rates of dispersal, but these have not been estimated per unit area occupied by a population.

The term N_i/a_i in equation (7.3B) can be interpreted as a density. That is, for a population of species *i* living in a specified area of size a_i, population pressure is simply density times the per capita rate of increase. But this is just another way of writing the rate of change in density, that is, if $D_i = N_i/a_i$, then population pressure is simply dD_i/dt. Thus, the theory of population dynamics is, in Lotka's view, a statistical theory isomorphic in its broad outlines with statistical mechanics. There are important differences. Most notable is the fact that individual "particles" in a population are organisms, and each organism is unique. In some populations this variability among individuals may itself be an important part of population dynamics, particularly if N_i is small (see also my brief discussion in chap. 2). In small populations, changes in population density depend on which particular individuals are able to breed each breeding season, and which survive from one season to the next. Relatively minor accidents may profoundly affect the growth rate. Thus, population dynamics are closely tied to chance events (e.g., Ewens et al. 1987). However, as N_i gets large, this variation among individual organisms becomes less important since the net rate of per capita change is averaged over more and more individuals. Hence, Lotka's isomorphism is best viewed at the largest spatial scales possible. In the next section, I examine a simple mechanistic model where N_i is the total population size.

An additional complication with Lotka's law of intensity is that, for many kinds of species, there is a significant feedback effect of population density on the per capita rate of change. That is, $\partial f_i/\partial D_i < 0$ for densities above some minimum density, say D_i', so that as density increases above D_i', the per capita rate of change decreases. Holt et al. (1997) made the simplifying assumption that density dependence does not vary in space. If this is so, then we can replace f_i with its density-independent component (Holt et al. 1997). This allows us to examine the effect of environmental parameters, $P_1, P_2, \ldots, P_n$, on density-independent population growth. Let ζ_i represent this density-independent component of population growth, so that

$$\max(f_i) = \zeta_i(P_1, P_2, \ldots, P_n).$$

We can now rewrite Lotka's intensity law as

$$\rho_i = \frac{N_i}{a_i}\zeta_i. \tag{7.4}$$

It is interesting to note that geographic range area can also be considered a macroscopic variable. That is,

$$a_i = \frac{N_i}{\rho_i} \zeta_i \tag{7.5}$$

is another way of examining the implications of Lotka's intensity law. This is very close to the relationship between distribution and abundance described in chapter 6. Equation (7.5) implies that there is a linear relationship between the logarithm of geographic range area and the logarithm of average population density. This is approximately what we see in regional assemblages (e.g., Hanski 1982a; Hanski, Kouki, and Halkka 1993), but is less evident in plots of the relationship when entire geographic ranges of each species are included. These plots seem to have boundaries (fig. 7.1), so that all possible combinations are not obtained; but there is no linear relationship as implied by equation (7.5). However, recall that f_i in equation (7.1) varies among species. In particular, the effects of environmental parameters (the P's in eq. [7.1]) on per capita rates of change should be expected to vary considerably from one genus to the next. Congeneric species for which these parameters might be similar are on the same line (fig. 7.2). The complexity seen in figure 7.1 derives from the heterogeneity of factors influencing the different species included.

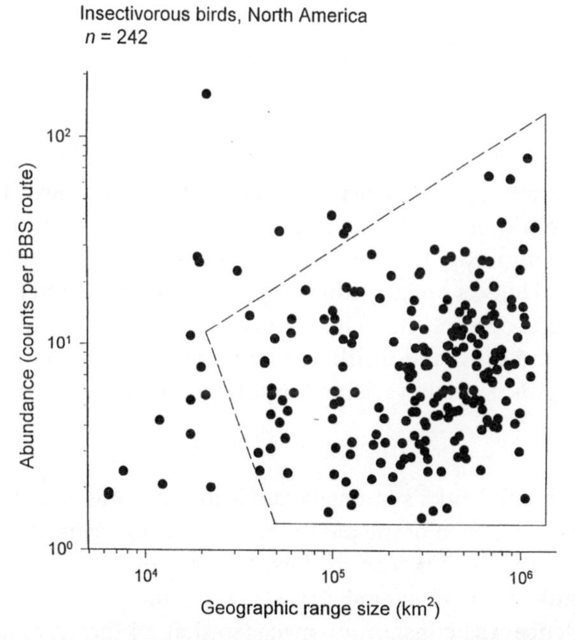

Fig. 7.1 Relationship between geographic range size and abundance for species of North American terrestrial birds. Boundaries are those hypothesized to exist in Brown and Maurer 1987.

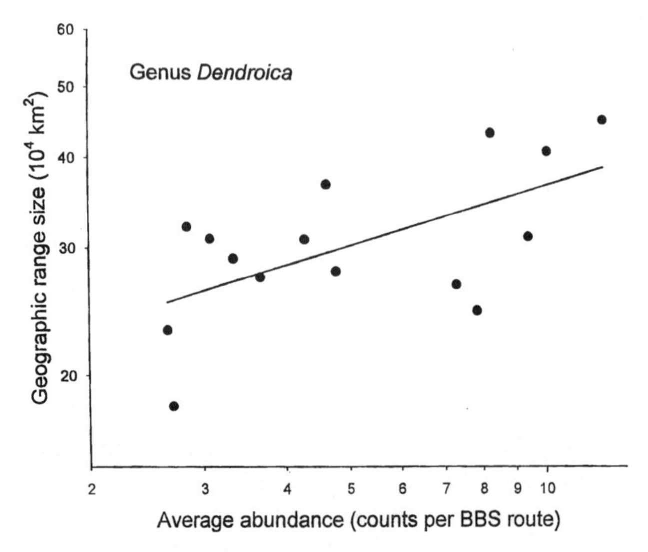

Fig. 7.2 Relationship between breeding geographic range size and abundance for species of wood warblers in the genus *Dendroica.* Most of these species breed in forested habitats in North America and are exclusively insectivorous. Virtually all species are migratory, with most wintering in Central and South America or the islands of the Caribbean.

A final point about Lotka's grand vision is that his explicit representation of the per capita rate of change as a function of environmental conditions (the P's in eq. [7.1]) foreshadowed the concept of a niche. The multidimensional nature of a niche is explicit in Lotka's formulation, since it is assumed that there are multiple environmental parameters that affect population dynamics. More important, though, are the mechanisms by which those environmental parameters control population dynamics: the complexity inherent in dealing with an unknown number of environmental effects on individual organisms is reduced to at most four demographic processes, namely, immigration and emigration in open populations and birth and death in all populations. The output of those four processes is a single variable: population abundance or density.

A Mechanistic Example

The problem with Lotka's intensity law is that it is not clear how exactly to determine what each of the parameters means in terms of known population processes. In this section, I explore this using a relatively simple approach outlined by Holt et al. (1997). Their model assumed that density dependence was constant in space, so that all that varied from one population to the next was the density-independent component of population growth. Furthermore, they assumed that there was a relatively low rate of migration among subpopulations, so that the demographic effects of immigration and emigration could be ignored. These two simplifying assumptions make the model relatively straightforward.

We begin with a slightly modified version of Holt et al.'s (1997) logistic equation for population growth. Let $N(x)$ be the population density of a species at some point in space (x). The density-independent component of per capita population change, $r(x)$, also varies with space. We assume that u, the density-dependent term, is a constant. At any point in space, the rate of change in population abundance is

$$\frac{1}{N(x)}\frac{dN(x)}{dt} = r(x) - uN(x).$$

At location x, the equilibrium density is

$$K(x) = \frac{r(x)}{u}.$$

Suppose that $r(x)$ is distributed across space in a Gaussian (normal) distribution with a spatial variance (this determines the width of the normal distribution) of σ_r^2. Figure 7.3 shows the temporal behavior of such a system. The spatial pattern in density converges on a normal distribution

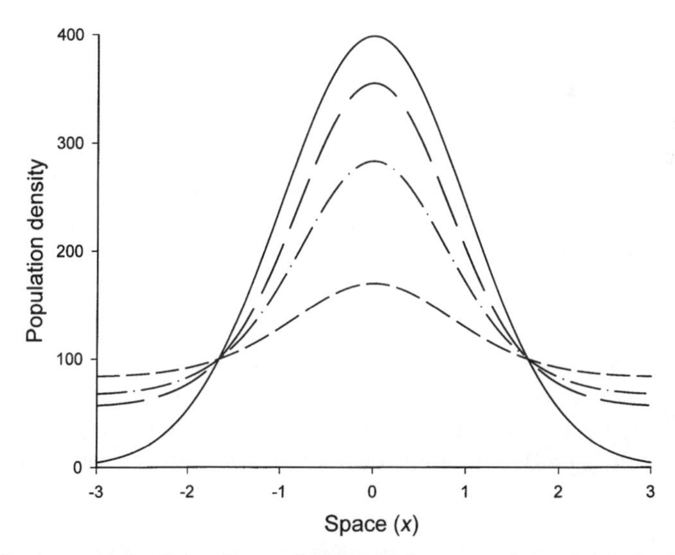

Fig. 7.3 An example of the changes in the spatial pattern of population density for a species initially introduced at equal densities across a landscape. Density dependence is constant across space and density independence is site specific. Time progresses from the distribution with the lowest mode to the distribution with the highest mode, which is also the equilibrium spatial pattern.

from any starting conditions (assuming all sites have a positive initial density).

The spatial variation in carrying capacity, σ_K^2, is related to the variation in density independence by the following relationship

$$\sigma_K = \frac{r_T}{u}\sigma_r,$$

where r_T is the sum of the $r(x)$'s across space. This relationship has two important implications. First, since σ_K is a measure of the size of the range of a species, range size will be larger in species with a lower intensity of density dependence, given the same maximum rate of increase for the entire population (summed across sites). In addition, the maximum rate of increase of the entire population of a species has a positive effect on range size. This expression, then, defines a relationship between the range of a species and its demographic characteristics.

The conclusions in the last paragraph hold for both one- and two-dimensional ranges. Considering two-dimensional ranges allows us to develop a mechanistic analogue for Lotka's intensity law. Assuming a two-dimensional Gaussian shape for a geographic range, we can approximate its area as the product of the spatial standard deviations along the major and minor axes. This leads to a relationship between the area of the range, a_K, and the area covered by the spatial variation in density-independent growth, a_r. Letting N_T be the total population size across a species' range, we have

$$a_K = \frac{r_T}{u}N_T a_r.$$

Letting $\rho = u/a_r$ represent Lotka's population pressure term, we reformulate the intensity law as

$$a_K\rho = N_T r_T.$$

Note that population pressure for this model increases with increasing density dependence. More importantly, it should be possible, for species that meet these assumptions, to predict the size of their geographic ranges directly from their demographic characteristics, allowing for a test of the model.

Brown's Principle

The version of Lotka's intensity law in equation (7.4) provides a unique opportunity to quantify Brown's (1984) ideas regarding the relationship between the abundance of a species and its distribution that I discussed in the last chapter. The quantification that I propose produces a testable prediction from Brown's basic premises. In this section I show the relationship between Brown's principle and Lotka's intensity law and describe a test of a prediction derived from the principle.

The basic idea in Brown's argument is that there are combinations of ecological conditions that are most favorable for a species, and that these combinations occur in limited regions of a species' geographic range (Brown 1984; Brown, Mehlman, and Stevens 1995). What does "most favorable" mean? Typically, favorability of a set of environmental conditions for a species, often referred to as habitat suitability, is measured by the average fitness that an individual can expect when exposed to those conditions. If we ignore density dependence, this is what the function ζ_i in equation (7.4) describes. Thus, if $P_1^*, P_2^*, \ldots, P_n^*$ represent the set of "optimal" environmental conditions for a species, then maximum favorability under these conditions is

$$\max \zeta_i \equiv \zeta_i(P_1^*, P_2^*, \ldots, P_n^*), \tag{7.6}$$

where $\max \zeta_i$ denotes the maximum fitness for species i. Furthermore, Brown's idea assumes that there is a monotonic ranking of sites by their values for the parameters P, such that for any pair of sites, those with values more similar to the set $\{P_1^*, P_2^*, \ldots, P_n^*\}$ will have values of ζ_i closer to $\max \zeta_i$ (Brown 1984; Brown, Mehlman, and Stevens 1995). Here I have avoided giving an explicit definition of what "more similar" means, because to be any more precise, we would have to make specific assumptions regarding exactly what the function ζ looks like. Brown assumed that it was Gaussian, and this may be as good a first approximation as any; but for the purposes of this discussion, we do not need to specify what ζ looks like.

We are now in a position to give a more precise statement of Brown's principle using Lotka's intensity law. The principle is as follows:

> For any set of environmental conditions $\Phi = \{P_1^*, P_2^*, \ldots, P_n^*\}$ such that $\zeta_i \{P_1^*, P_2^*, \ldots, P_n^*\} = \max \zeta_i$, the density of species i, D_i^*, in any geographic regions that contain the conditions described by Φ, will be the maximum density.

I will consider the empirical interpretation of this statement in a moment. First, consider the following prediction. Since ρ_i, the population pressure, is the product of density times the maximum per capita rate of change (i.e., $\rho_i = D_i \zeta_i$), the geographic regions that contain the optimal set of environmental conditions should experience the highest population pressure. Before examining how to test this prediction, I want to go back and look at Brown's principle.

Initially it seems reasonable to expect that those regions that have the highest suitability will be those regions that have the highest density. However, there are some reasons to think that this might not always be the case. In particular, there are thought to be some biological conditions that might prevent this from happening (Van Horne 1983; Pulliam 1988). If, for example, the social system of a species allows for dominant individuals to monopolize the best resources, individuals of lower social status may be limited to relatively poor habitats. If there is a sufficiently large difference between the resources needed for individuals to survive and those needed for individuals to reproduce, it is possible that individuals could accumulate to such an extent in poor habitats that densities exceeded those in favorable habitats. Thus, density might be inversely related to habitat suitability. How often this kind of habitat selection occurs is unclear, and is something about which we need further information.

The relationship between habitat suitability and density, however, may be different on different scales. It may be that the kind of "source-sink" structure described in the previous paragraph may occur within landscapes, so that patches in these landscapes that act as sources (i.e., $f_i > 0$ in the patch) have lower densities than those that serve as sinks (i.e., $f_i < 0$ in the patch). But at a larger scale, conditions may vary from one landscape to the next such that some landscapes have a higher density of both good and bad patches when compared to other landscapes. When suitability and density are averaged over landscapes, the relationship between average fitness and density might be positive, even if it is negative within landscapes. Thus, in the best landscapes, environmental parameters are more favorable across the entire landscape, even though they vary among patches within landscapes. At the scale of the entire geographic range, the assumptions made by Brown's principle may be met, even if the assumption does not hold up within individual landscapes.

Recall that using Brown's principle I predicted that population pressure should be highest in those regions of a species' geographic range where density (and fitness) are highest. If population pressure were easy to measure, then it would be a simple study to measure the pressure in regions within a geographic range that supported different densities. Intuitively, population pressure would seem to be related to dispersal rate, so that the necessary measurements would be dispersal distances of individuals in different parts of the geographic range. The higher the population pressure, the farther individuals might move before settling. However, things are probably not that simple. It is possible to have situations where population pressure may be high, yet movements of individuals are restricted. For example, if habitat is sufficiently limited, individuals may remain where they are rather than dispersing, since there is nowhere to disperse to. The environment could be oversaturated, but individuals would simply not move.

A clear test of the prediction I derived from Brown's principle could be extremely difficult to obtain for existing populations. Consider the following thought experiment. Suppose we were able to identify where environmental conditions were best for a species by identifying high-density regions of its geographic range. If we were able to remove the species from its range instantaneously and then reintroduce the species at a number of random locations, we would predict that those regions identified as being high-density regions would fill faster than those regions identified as low-density regions because the high-density regions would have higher population pressure than the low-density regions (see also Holt et al. 1997). Although such an experiment would be questionable ethically, very similar experiments have been conducted inadvertently by the introduction of exotic species and naturally by colonization of continents by new species. An example is the European starling, introduced into North America in the late 1890s (Wing 1943). Starling populations expanded relatively slowly during the first decade after their release in New York. After about thirty years, they began to expand rapidly into the southeastern United States. Meanwhile, they moved relatively slowly northward through New England and Canada. The rate of expansion is clearly related to their current abundance (Maurer and Villard 1994). Starlings moved faster through those regions where they are now most abundant. In this case, the pressure of the population to expand was greatest in those regions which now seem to harbor the best set of ecological conditions for the species. Although anecdotal, this example of range expansion suggests that Brown's principle may prove to be an example of a large-scale generalization that can generate predictions regarding the outcome of continental-scale population dynamics.

A Thermodynamic Interpretation of ζ, the Maximum Per Capita Rate of Population Change

Before leaving this discussion of Lotka's intensity law, I want to revisit the isomorphism between the maximum per capita rate of change and temperature. Both quantities express the potential of the system to undergo its dynamics. For temperature, there is a direct thermodynamic effect on the system: higher temperature imparts a higher level of energy to individual molecules, which in turn move faster, leading to a change in the state of the system (either expansion of the volume of the gas or increased pressure). According to Lotka's intensity law, the density-independent component of the per capita rate of population change has a similar effect on population pressure: when it is high, there are many organisms vying for space and population pressure is accordingly high.

Despite these intriguing similarities, the maximum per capita rate of change is a very different kind of quantity from temperature. Temperature in the ideal gas law signifies that a distinct energy gradient exists from the outside of the system containing an ideal gas to the inside, and that energy flows to and from the inside of the system (though the net energy flow could be zero). These energy flows come in the form of energy transformations, so that energy of one kind is transformed into energy of another. Population dynamics has not traditionally been viewed as an energy transformation process. Rather, it has been viewed more as an accounting of how many individuals are alive, how many are giving birth, how many are dying, and so forth.

This traditional view of population processes has sometimes obscured the fact that in order for reproduction to occur, energy transformations must be continually carried out as individual organisms obtain energy and use it for maintenance and reproduction. These energy transformations occur through biochemical reactions carried out within organ systems of individual organisms. Without a continual flow of energy through organisms, no population-level process could occur. Hence, population change can be viewed as a thermodynamic process as well as an accounting process that describes the fate of individual organisms. These alternative descriptions of populations should be complementary, so that population concepts like fitness should have alternative thermodynamic formulations.

Despite the expectation that population dynamics could be describable as an energetic process, relatively little progress has been made in producing an appropriate formulation. Even relatively simple systems (e.g., two competing species) give rise to rather complicated descriptions when attempts are made to represent population processes in terms of energy flows (e.g., Maurer 1990b; Yodzis and Innes 1992). Yodzis and Innes (1992) pointed out that it is a close-to-impossible task to detail all of the possible factors that must be accounted for to describe the detailed thermodynamic behavior of a population system. The way around this complexity is to develop parameterizations that summarize (given appropriate assumptions) much of the complexity. There are a number of ways to do this (see Yodzis and Innes 1992; Perrin and Sibly 1993; Brown, Marquet, and Taper 1993), but each has its drawbacks (Kozłowski 1996; Brown, Taper, and Marquet 1996). Nevertheless, in general, it is safe to say that thermodynamics should eventually play a major role in helping us understand the nature of population dynamics.

Geographic Range Structure

If Lotka was correct in his general way of viewing biological populations, it follows that we might see evidence of population processes in the spatial and temporal patterns of populations. This expectation, of course, has motivated most work on spatial and temporal population dynamics and is quite independent of Lotka's approach. But my emphasis on the statistical nature of Lotka's ideas suggests that the more expanded the spatial scale at which we view a population, the more likely we are to see some kind of significant pattern that has some repeatability or regularity among species. Obviously, the largest spatial scale at which it is possible to study populations is the geographic range of a species.

Geographic ranges are not randomly constructed geometric objects distributed without pattern across geographic space. Rather, they are structured in a nonrandom fashion. We do not yet have a theory that can readily predict from first principles what these structures should look like. Rapoport (1982) and Brown (1995) have provided some initial forays in this direction, and the ideas discussed above may eventually be fleshed out sufficiently to provide such predictions. In the mean time, it is necessary to explore just what patterns exist and how these patterns might be tied to the kind of theory envisioned in the discussion of the previous section.

Patterns of Population Stability and Abundance across a Geographic Range

In chapter 6 I introduced a pattern that I claimed was rather widespread among taxa, that is, that a species tends to be most abundant near the center of its geographic range, and its abundance declines with the distance away from this peak. Many species have multiple peaks, but these peaks are always surrounded by regions where density declines gradually in all directions away from the peak until either the geographic range boundary is reached or another peak is encountered (see Root 1988c and Price, Droege, and Price 1995 for bird examples; Hengeveld 1990 for examples from other taxa).

There is more to this pattern than the regular decline in abundance away from abundance centers. Curnutt, Pimm, and I (1996) showed that abundance varied relatively more over time in peripheral populations than in central, high-abundance populations for several species of birds. This implies that central populations are more stable over time and less prone to extinction than peripheral populations. One can estimate the stability of a population by examining the average per capita rate of change in the population over time. If it is close to zero, the population is relatively stable; if it is positive, the population is increasing; if it is negative, the population is decreasing. Furthermore, the likelihood of extinction of a population is related to the variance in the per capita rate of change of the population (Goodman 1987).

The stability of populations across a species' geographic range varies in a striking fashion. I have plotted patterns for the grasshopper sparrow in figure 7.4. Note that virtually all of the high-abundance sites have per capita rates of growth close to zero, while the low-abundance sites vary considerably in their per capita rates of change (fig. 7.4A). This means

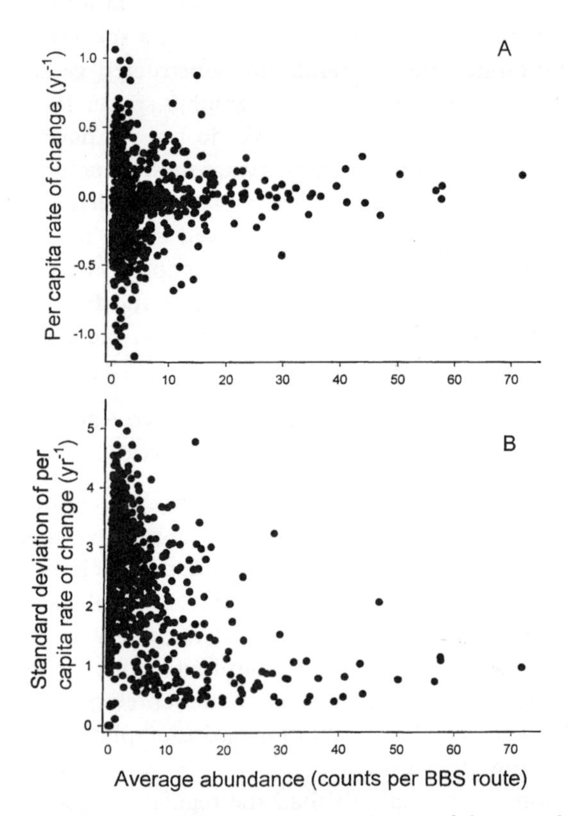

Fig. 7.4 *A,* relationship between the average per capita rate of change and average abundance for populations of grasshopper sparrows estimated from the BBS. Averages were calculated over time for all BBS routes that were censused at least ten times in the past 30 years. *B,* relationship between standard deviation of per capita rate of change and average abundance for grasshopper sparrows on the same BBS routes as in panel A. Note that high-abundance sites generally have rates of change close to zero and have relatively small variances for that rate of change. This implies that these high-abundance sites are relatively stable over time. Furthermore, low-abundance sites tend to have high rates of change and high variances of that rate, implying that these peripheral sites are less stable over time.

that high-abundance sites maintain relatively stable populations from one year to the next, and whatever changes in abundance occur are relatively small. Sites with low abundances, on the other hand, can be changing drastically from one year to the next, and the net change may be either positive or negative. Furthermore, notice that the high-abundance sites have relatively small variances for the per capita rate of change compared to the low-abundance sites (fig. 7.4B). This implies that low-abundance sites have a higher likelihood, on the average, of local extinction compared to high-abundance sites.

The estimates of per capita rates of change in figure 7.4 can be considered estimates of $f_i(P_1, P_2, \ldots, P_n)$ in equation (7.1). Each point is an estimate of this quantity for a given set of P's since each point represents a different location with a different combination of environmental conditions. The high-abundance sites appear to be highly regulated since they appear to be the most stable sites, whereas the low-abundance sites seem to experience little regulation at all. Recall that Lotka's intensity law predicts that population pressure should be highest in these high-abundance sites. If this is true, then the pattern exhibited in figure 7.4 implies that high population pressure occurs in regions of high population stability within a species' geographic range.

If population pressure is high in high-abundance sites, then it seems that there would be a continual excess of individuals being produced in these sites. But since these populations are also stable, it implies that excess individuals produced in high-abundance sites must go elsewhere to find suitable habitat. This further implies that there should be a net movement of individuals from the central, high-abundance sites in a geographic range toward the periphery. These movements may be relatively slow, and could easily occur across many generations. This continual flow of individuals into peripheral sites from central sites could contribute to the high variance in per capita rates of population change in peripheral sites. With strong density dependence, it is possible that environmental variation at high-abundance sites might actually drive fluctuations at the periphery of the range.

To my knowledge, no one has actually measured such a net flow of individuals from central, high-abundance populations to peripheral, low-abundance populations within the geographic range of a species. There might be genetic consequences to such a migration that in principle could be measured. An evolutionary consequence of such a process may be niche conservatism (e.g., Ricklefs and Latham 1992), where the niche of a species does not evolve over time because any adaptation to conditions different from those in source habitats cannot proceed due to population flow from the sources (Holt 1996).

Geographic Range Fragmentation

In the last section, a picture of the spatial dynamics of populations of a species across its geographic range emerged that emphasized the dynamical importance of the high-abundance sites in a geographic range. The stable central populations of a geographic range presumably provide a net flow of individuals away from these central populations toward peripheral sites, although there are few, if any, data on whether there is in fact such a differential migration. Furthermore, it is unclear how such a process might be maintained in species that undergo seasonal migrations to and from their breeding grounds. There are some important implications of this picture of geographic range structure. In this section, I explore how geographic range structure varies among species and what consequences this variation has for the evolution of biological diversity.

In order to understand the ecological mechanisms that underlie the demographic structure of geographic ranges and how those mechanisms might vary from one species to the next, consider the following simple example. Imagine two related species that have similar ecological characteristics and tolerances, with a single exception. Suppose one species is relatively generalized in the foods that it can eat and the other species will eat only a few foods. The species that uses more foods should be found in more diverse places, because in those places it should be able to find at least some appropriate kinds of food. The other species would necessarily be limited to only those places where it could find the few foods it was capable of eating. The range of foods that a species eats tells us little about the population dynamics of that species in any local site; we need to know something about the amount and quality of those foods at the local site before we can predict what the birth and death rates at that site will be. However, all else being equal, and assuming that food is what limits the populations of the two species, we could predict that the species that can eat a greater variety of foods will occur at more places than the one that eats fewer foods. Thus, the species with a more diverse diet will have a larger geographic range than the species with a narrow diet. This is a very simplified version of Brown's (1984) hypothesis regarding differences in geographic range size between ecologically generalized species and ecological specialists.

In the case of these two hypothetical species limited only by food, knowing that the geographic range of the more generalized species is larger does not provide any information regarding the *shape* of the geographic range. Geographic ranges do not have simple geometric shapes. The geometric figure described by the average geographic range boundary of a species is highly convoluted (see examples in Root 1988c; Maurer

1994; Price, Droege, and Price 1995). Such a figure can be considered to be "fragmented" because there are often small "islands" produced along the edge of a geographic range where populations are expected to be found. Does this pattern of fragmentation differ between species that have large geographic ranges and those that have small ranges? On the one hand, a generalized species, such as the hypothetical species that can eat a wide variety of foods, might be able to penetrate deeply into areas of inhospitable habitat and maintain small populations in these areas. If such extensions were common, then such a species might have a more jagged range boundary than a more specialized species. On the other hand, a species that is generalized may be able to effectively fill all habitats near the periphery of its range, so that its range boundary may be relatively smooth with populations being found relatively continuously along the boundary. Thus, there is no theoretical reason that a large range should have either a smoother or a more jagged shape than a small range.

Returning to the idea of demographic structure of geographic ranges, I have argued that there is no reason to expect that the demographic processes that underlie geographic range shape should produce a more or less jagged range shape for widespread species (that have large geographic ranges and can use or tolerate many different sets of environmental conditions) when compared to narrowly distributed ones. Should we even expect differences in geographic range structure at all between widespread and narrowly distributed species? Brown (1984) argued that we should. In the last section we found that, within a single species, there was considerable variation in abundance and stability of local populations. That variation had to do with Brown's principle. The more closely the environmental conditions of a local site approached the optimal set for a species, the more dense and stable the local population that inhabited the site. Brown (1984) argued that species have significantly different optimal environmental conditions. For some species, the optimal conditions occurred widely across geographic space; for others, those conditions were only rarely met. How do these patterns of spatial variation in optimal conditions vary among species?

Phil Nott and I recently showed that among 242 species of insectivorous birds in North America, those with large geographic ranges had smoother range boundaries than those with small geographic ranges (Maurer and Nott 1998). Thus, widespread species are able to fill all habitats that occur within the range of their maximum ecological tolerances. So Brown (1984) was right: there are significant differences in the way that widespread and narrowly distributed species react demographically to environmental variation. Narrowly distributed species have more

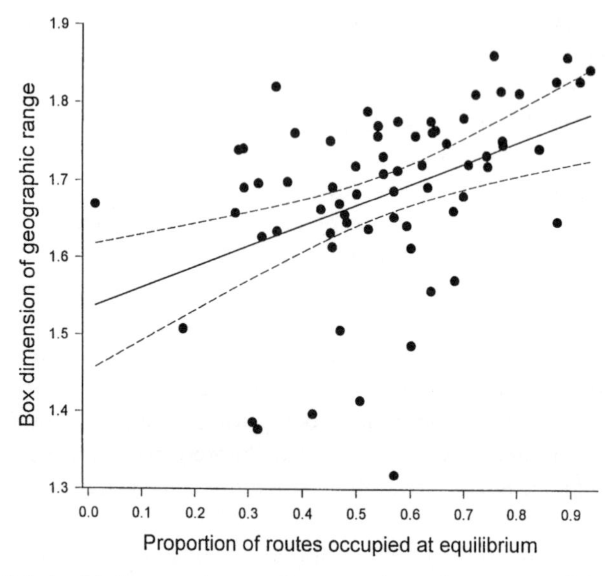

Fig. 7.5 Relationship between degree of fragmentation and occupancy of sites within the geographic range for species of insectivorous birds in North America. Note that box dimension is inversely related to fragmentation, so that species with low box dimensions have relatively fragmented ranges.

populations in contact with conditions that can be considered extreme for the species by virtue of their small geographic ranges, and this effect is exaggerated by the fact that species with small geographic ranges also have more-fragmented ranges. Nott and I went further to show that narrowly distributed species with fragmented geographic ranges occupy a smaller portion of their geographic ranges (fig. 7.5). Local populations of these species, particularly at the species' range boundary, are frequently becoming extinct and must subsequently be recolonized. This happens more often for the narrowly distributed species than the widespread species.

Geographic Range Structure and Extinction

There are some important consequences for the way we view extinction that follow from these patterns of variation in demographic structure of geographic ranges within and among species. Extinction can occur at many scales. Within a landscape, a population may perish within a single habitat patch, but persist in the landscape through recolonization from patches with viable populations. A species may become extinct in all patches within a landscape, but persist within its geographic range because it maintains viable metapopulations within at least some landscapes within its geographic range. Some landscapes may produce an abundance of individuals, acting essentially as source landscapes for colonies in new landscapes with appropriate habitat from which the species is missing. A species that is unable to persist in any landscape will face global extinction.

The nested hierarchy of extinction processes described in the previous paragraph must vary among widespread and narrowly distributed species. In my opinion, this variation is quantitative in nature rather than qualitative; that is, I do not think the processes of extinction operate any differently in widespread species versus rare ones. From the last section we learned that rare species have small, fragmented geographic ranges, and they occupy fewer local sites within these ranges. The local populations of these species are on the average smaller; hence, each local population has a higher likelihood of extinction than a closely related widespread species. Fewer local sites are occupied within any landscape, and consequently it is more likely that a rare species will become extinct within a given landscape. Furthermore, there should be fewer landscapes in a rare species' geographic range within which it can persist for long periods of time; hence, the occupiable fraction of its geographic range is smaller than that of a widespread species. Having a smaller fraction of landscapes that can serve as sources for landscapes where extinctions have occurred makes a narrowly distributed species more susceptible to global extinction than a species with a large geographic range.

We can tentatively conclude, then, that species differ in susceptibility to extinction primarily because they vary in the structure of their geographic ranges. There may be many reasons why a species might have a particular geographic range structure. Gaston (1994), for example, listed a number of ecological attributes that predispose a species to be rare. Whatever the reason for rarity, the net impact on spatial population dynamics of those ecological attributes that determine the degree of rarity is to produce characteristic patterns of geographic range structure such as those described above. Thus, geographic range structure is a macroscopic property of a species that is determined by the innumerable details of its natural history as they are played out across a continent. Accidents of history may form dispersal barriers that modify the structure of a geographic range, but it may be possible to understand such barriers in a mechanistic fashion by focusing on their effects on geographic variation in the demography of species.

The search for general principles governing geographic range structure will complement the findings of more intensive local studies by providing a clearer picture of the context in which those studies are done. The patterns discussed in this chapter suggest that such general principles exist. Whether they ultimately take the form of Lotka's law of intensity has yet to be determined, but his development of a population law that is isomorphic with the statistical mechanics of simple physical systems is intriguing. Very clearly, what needs to be done next is to obtain data sets on the geographical distribution of abundances for groups of organisms other than birds. Only with such data sets in hand can we begin to address the generality of the patterns discussed in this chapter.

CHAPTER EIGHT

Geographic Assembly of Local Communities

> Lower-level constraints—essentially intrinsic dispositions—are generative and propose possible results, while higher-level constraints are regulatory and dispose. Upper-level constraints dominate the activity and productivity of focal-level processes. This means . . . that individual upper-level perturbations are powerful in the sense that they simultaneously and globally affect all subsystems within the one where the perturbation originated. In contrast, the activities of these lower-level entities are weak because no one of them can have that kind of cogent, global influence upon the supersystem within which they exist.
>
> S. N. Salthe (1985)

The passage quoted above is part of an attempt to argue that causes can come to a system both from the parts of which the system is made and from the larger system within which the system itself resides (see chap. 2). The second kind of cause referred to by Salthe can be called "upper-level," as opposed to "lower-level," causation. In "traditional" community ecology, the focus has been on lower-level causation. The implicit assumption has been that in order to understand the structure of a community, it is necessary to understand the mechanics of interactions among the species that make up the community. Upper-level causation is often viewed in terms of perturbation effects. Interannual variation in weather, for example, can have a profound impact on the outcome of species' interactions, and can even reverse competitive relationships. Predicting such kinds of environmental variation is difficult or impossible. Furthermore, community ecologists cannot all expect to become meteorologists in order to predict the outcome of their field experiments or to understand their observations.

Recall from Salthe's quote above that upper-level causes are powerful in the sense that they affect all lower-level entities that they comprise. The same cannot be said of lower-level causes. Because such causes originate from parts of larger wholes, they may or may not affect the whole because such causes need not influence any of the other parts (though in some cases other parts might be affected indirectly). In terms of community ecology, this means that the outcome of a competitive relationship in a local community could conceivably have little effect on the landscape in which that community resides. However, disturbances that affect the entire landscape will clearly have the potential to affect the outcome of any particular species interaction.

Up to this point, I have referred to upper-level causes that are more or less random with respect to any particular community, such as large-scale disturbances. Some have argued that disturbances are common enough in most communities that we can never expect any kind of structure to emerge. This viewpoint is an extension of Gleason's concept of plant associations that I discussed in chapter 1. As persuasive as such a view might be, it ignores an important source of upper-level causation that is rarely considered in discussions of community structure, at least not explicitly. This source is the effect that phylogenetic history has on a community.

One of the consequences of the formation of a new species is the formation of a discrete gene pool. As this set of genes interacts with the environment during the ontogeny of each individual of the new species, a unique set of ecological characteristics arises. Individuals within species are nearly always more similar to one another in the details of their ecology than they are to other species. Since the geographic range structure of the species arises from the demographic consequences of these ecological details, this unique structure should be identifiable with the use of appropriate measures. This suggests that an important source of upper-level constraint is imposed upon an assemblage of species at any scale by the individual evolutionary histories of the constituent species (Brooks 1985; Brooks and McLennan 1991).

This oft-ignored component of upper-level causation can serve as a basis for looking for order that might exist within communities at different scales. It leads to new explanations for well-established large-scale synecological patterns. It can also explain somewhat anomalous observations regarding geographic variation in habitat associations. In conjunction with the processes documented by experimental ecology, it can be used to build a more robust set of community assembly rules for the local-scale communities which have been so problematic in the past.

Statistical Relationships between Synecology and Autecology

As I have argued from the outset, random processes are seldom truly random. In a system as complex as a community, there are so many events involved in the dynamics that it is essentially impossible to make precise predictions about outcomes. Despite this, communities persist through time, and though they are constantly changing, it is often possible to recognize at least some level of organization. The environment in which a community exists and the evolutionary histories of its constituent species provide ample opportunity for upper-level control of community structure, and it may be possible to understand these causes by conducting the kind of input-output analysis that I described at the end of chapter 5.

The explanatory power of detailed analyses of communities at local spatial scales has its limitations (chap. 3). This has led some ecologists to advocate that we study autecology and deemphasize synecology. Autecology is the study of how individual organisms relate to their environment. To understand population dynamics mechanistically, it is necessary to describe the fate of each individual that enters the population and leaves it. This means that each individual should be followed from its birth or immigration through its reproductive activities until it leaves the population by dying or emigrating. If a small enough population is studied, this is not an unrealistic task (Andrewartha and Birch 1984). As the spatial scale over which a population is examined increases, the likelihood that the task can be completed decreases rapidly. It is at this larger scale, however, that I have argued we must look to find the kinds of principles that will help us make further progress in understanding ecological systems (chaps. 6 and 7).

Synecology is the study of the interactions among organisms. It is at this level of description that ecologists had hoped that general laws would emerge. Presumably, there is a relationship between the autecological approach and synecology. Some have been suggested, but there are few that have endured critical analysis.

Frank Preston's statistical approach to synecology was not uncommon during the middle part of this century. C. B. Williams, R. A. Fisher, L. R. Taylor, J. G. Skellam, Preston, and many others built a tradition where assumptions regarding organismal behavior were translated into characteristic statistical distributions with parameters that could be measured experimentally (see chap. 2 for a brief discussion of some of these statistical theories). The problem with statistical theories was that they were often mathematically intractable to many ecologists. Calculating parameters was difficult if not impossible without computers. Even if parameter estimates were obtained, interpretation was often obscure or difficult, and of limited usefulness in light of the assumptions that were made in order to derive the theories in the first place. Statistical theories gave way in the mid- to late 1960s to the experimental approaches of J. L. Harper, R. T. Paine, J. H. Connell, and others. With this emphasis on experiments came a new reductionist philosophy that incorporated linear thinking. I have already described the limitations of such thinking when it is applied to ecological systems.

Throughout this book I have advocated revisiting the statistical approach to ecology. Problems with many past attempts to construct statistical theories of ecology have centered around unrealistic assumptions made for the purpose of rendering the mathematics tractable. Yet such assumptions often lack solid empirical support, as was the case with Preston's derivation of the species-area relationship. In this section I discuss "synecological" patterns, such as the species-area relationship, from the perspectives developed in the last two chapters. Macroecology increased our understanding of how abundance is distributed in geographic space. The concept of geographic range structure discussed in chapter 7 builds upon this understanding to develop mechanistic descriptions of how this abundance distribution arises from the autecology of individual organisms. The clumped nature of abundance within a geographic range is an empirical fact. It will be seen in the ensuing discussion that using this observation as a fundamental premise gives rise to new explanations for patterns like the species-area relationship.

Simulating Geographic Species Assemblages

Formal mathematical treatments of the statistical mechanics of species assemblages have been attempted in the past. As I pointed out in chapter 2, these theories were of limited usefulness because they made too many unrealistic assumptions. Similar approaches could conceivably be constructed using more realistic assumptions, but the mathematics required to do so would probably be prohibitively complicated. However, with advances in computer technology being what they are, it is often possible to obtain numerical solutions to complicated mathematical problems that cannot be solved analytically (i.e., a set of equations cannot be obtained that provides the solution to the problem).

Numerical solutions to complicated mathematical problems have a long history in ecology. Because ecological data are so variable, it is often difficult to differentiate signal from noise. One approach to this problem has been to develop simple models that generate noise alone, and then compare the pattern generated by noise to a pattern proposed to be determined by some process of interest (Gotelli and Graves 1996). These simple models are called "null models" because they attempt to model a process that produces an ecological pattern without invoking specific ecological mechanisms. The pattern produced by the null model can then be compared to the empirical pattern. In this way, the role of a particular mechanism can be evaluated against an alternative process that does not include that mechanism.

In the following sections I use a null model approach to show how the addition of realistic constraints in the modeling process allows one to evaluate the role of top-down causal factors in generating well-known synecological patterns. The three patterns I evaluate are the species-area relationship, the relationship between distribution and abundance, and the pattern of nestedness among species distributed across space. Each of these patterns was intensively studied in the past and a variety of different attempts were made to generate causal explanations for them.

The Species-Area Relationship

One of the earliest patterns that arose from synecology was the species-area curve (Rosenzweig 1995). The basic observation was that as the area surveyed increased, the number of species observed within the area also increased. This was as true for continental samples as it was for island systems (Rosenzweig 1995). There appeared to be a regularity to this relationship. Most commonly it has been expressed as a power law, that is, if S is the number of species and A the amount of area surveyed, or the size of an island, then

$$S = cA^z. \qquad \textbf{(8.1)}$$

How did such a pattern arise from the autecology of the numerous individual organisms that composed the populations contained within an area of a given size?

An answer was provided by Preston (1962a,b). He assumed that the intimate details of how every individual organism underwent its life history led to characteristic spatial distributions of populations. This simplifying assumption allowed him to derive a statistical distribution that described variation in abundances among species. This distribution, of course, was the famous lognormal distribution. After deriving this distribution, he deduced that the species-area relationship given by equation (8.1) followed directly from it. In fact, he argued that if the distribution of abundances among species fulfilled certain assumptions (the so-called canonical lognormal distribution), then the value of z in equation (8.1) should be around 0.26.

Unfortunately, the conclusions derived from Preston's (1962a,b) theory have not fared well. Empirical evidence is often used to argue that canonical lognormal distributions are rare. Furthermore, the predicted value of z that Preston deduced has not held up to empirical evaluation. For example, Connor and McCoy (1979) showed that there was a large amount of variation in estimates of z calculated from species assemblages at different geographic locations. They concluded that no single value for this parameter was appropriate, as Preston's theory required.

Preston's ideas were applied mainly to island populations, but he also tried to explain a related continental pattern. On islands, the species-area relationship was applied to islands of different size. On a mainland, as the size of an area surveyed is increased, the number of species counted within that area increases. This is widely seen as a sampling phenomenon, that is, as area increases, rare species that are not likely to be found on small plots begin to appear in the successively larger areas being sampled, so the number of species must increase.

Gotelli and Graves (1996) reviewed four hypotheses regarding the species-area relationship in an island archipelago. Two of these hypotheses invoked random processes. The first argued that small islands are more likely to have disturbances that periodically cause extinctions on the island, thus reducing species diversity on them relative to larger islands (McGuinness 1984). Hence, on the average, species diversity will be lower on small islands than on comparable larger islands. A second hypothesis assumes that dispersal among islands is completely random, so that the only effect of island size is to act as a "target" (Coleman et al. 1982). The larger the target, the more species that are able to locate it; hence, larger islands should have more species than small ones simply by passive sampling alone. Note that both of these hypotheses appeal to different kinds of randomness.

Two additional hypotheses reviewed by Gotelli and Graves (1996) assume that differences between large and small islands are due to a more deterministic set of processes. The first of these assumes that larger islands have more kinds of habitats or resources than smaller ones, so that larger islands can support a higher diversity of species than small ones (Williamson 1981). The second, the most studied of all of these hypotheses, is the "equilibrium theory" of MacArthur and Wilson (1967). The basic idea of this hypothesis is that the number of species on an island is determined when rates of colonization and extinction on the island are balanced, and hence a steady state is achieved. Thus, the number of species remains constant, while there is a constant turnover of species as some become extinct and others colonize the island.

The "equilibrium" theory, or more accurately, the steady state theory, has been widely tested, with some evidence supporting it and other evidence contradicting it (Gotelli and Graves 1996). The other hypotheses, though having received less attention, also have some degree of support. However, after examining each hypothesis, it becomes clear that they are not mutually exclusive. That is, any particular data set can be consistent with more than one of these four hypotheses, and distinguishing among them is often difficult (see Gotelli and Graves 1996, table 8.1). Here, I examine how these hypotheses might be applied to continental species-area relationships.

Some kinds of habitats on continents are clearly islands, that is, they are discrete and widely separated. Such habitats include caves, mountain tops, and freshwater ponds. Most habitats, however, though patchy across a geographic region, are not isolated enough to be considered distinct islands. Often, habitats intergrade into one another across relatively large ecotones. Patches may also be connected to one another by corridors of similar habitat. Ecotones and corridors allow individuals to move relatively freely among patches, effectively creating large, loosely connected populations. At the continental scale, regional populations often appear to be nearly continuous, although this breaks down at smaller spatial scales (see, e.g., Brown 1995). The important point is that the lack of isolation of most habitats makes it unclear how hypotheses originally intended to explain the species-area relationship on islands might be modified to explain the species-area relationship on continents.

The hypotheses that assume that species-area curves result from random processes are most easily extended to the continental situation. The disturbance hypothesis suggests that smaller areas within a region are more likely to suffer disturbances than larger ones. Disturbances vary in extent, such that the distribution of disturbances often follows a "hollow curve." That is, there are many small disturbances and very few large ones. If this is generally true, then small areas that experience a disturbance will be more likely to have their entire area affected by the disturbance, while larger areas will have a higher probability of having only part of the area affected by a disturbance. A disturbance that affects an entire area will create a uniform habitat within that area, while one that only affects part of the area will create two or more distinct habitats. Larger areas will be more likely to comprise a patchwork of successional habitats than will smaller ones, hence larger areas will harbor more kinds of species. Note the similarity between this hypothesis and the habitat diversity hypothesis discussed in the next paragraph. The major difference between them is that the disturbance hypothesis assumes that habitat heterogeneity is generated only by disturbances. The second hypothesis that incorporates random processes is one of passive sampling. This hypothesis simply states that, within a geographic region, larger areas will be more likely to be found by more species than smaller ones, even if all species are relatively uniformly distributed at the large scale. If settlement is completely random, then the smaller the area, the more likely individuals of a few species will miss it, hence, a decrease in diversity is expected as the area sampled gets smaller.

Since habitats are not continuous in space, increasing the surveyed area of a geographic region will increase the number of habitats encountered in the sample region. Thus, the habitat diversity hypothesis applies readily to continental systems (Williamson 1981). Sites of a given size are heterogeneous, according to this hypothesis, because a number of processes ensure them to be so. Since climate, geology, and topography are not uniform across geographic space, natural boundaries are formed by variation in these physical factors, leading to the expectation that species diversity varies across space. Since larger areas have a greater likelihood of crossing these natural boundaries, more habitats, and hence more species, are expected. The similarity between this hypothesis and the disturbance hypothesis suggests that both could be included in a single "habitat heterogeneity" hypothesis. The difference between the habitat heterogeneity model and the passive sampling model is that the former assumes local sites vary across space, while the latter does not.

Extending the steady state model to continents is not straightforward. Since most populations are continuous across space, local extinction is relatively rare, and will often be followed by recolonization. But how quickly the recolonization occurs will depend on where within a particular species' geographic range a site is located. Sites on the periphery of a geographic range have a lower probability of being colonized than those nearer to a high-abundance region (chap. 7). Hence, species do not have equal access to all sites in the geographic region. That is, the degree of isolation of a particular site is different for each species, and depends on the ecological characteristics for each species that determine the geographic range boundary (Root 1988a,b). Thus, species cannot be considered to have equal probabilities of extinction at a given site; some species will have much higher probabilities of local extinction than others. The same is true of colonization. The likelihood of successful colonization of a given site for each species varies depending on where the site is located within each species' geographic range. Rather than a steady state where species are constantly becoming extinct and recolonizing the site, the expected pattern is that some species would persist indefinitely at relatively high densities; other species would be unable to establish permanent populations, persisting for short periods of time before becoming extinct; and some species might be intermediate between these extremes (Williamson 1981). There is no reason to expect such a process to reach a steady state. The problem with applying the steady state model to continents, then, is that there is considerable spatial heterogeneity among species in the demographic processes that determine their ability to colonize and persist across the continent or geographic region being considered.

An alternative mechanistic model to the steady state model is based on the concepts of geographic range structure developed in chapter 7. Recall that the abundance of a species varies considerably across its geographic range, with the highest abundance in centrally located sites, and abundance declining away from these centers toward the range periphery. Population stability varies along this density gradient, so that peripheral populations are constantly becoming extinct and being recolonized. Hence, colonization and extinction dynamics governing the diversity at local sites are caused as much by heterogeneity *among species* as by habitat differences among sites. Furthermore, it is possible, knowing the geographic distributions of species, to predict which species are more likely to persist at a given site over time, and which are more likely to go extinct. Species diversity will increase with increasing area because larger areas will contain portions of more species' geographic ranges than smaller areas.

None of the hypotheses discussed for continents makes any explicit predictions about what the values of the power relationship should be. The steady state model has been extended to suggest that z should be around 0.25 (e.g., Sugihara 1980), but it is not clear that such extensions apply to continents. Rosenzweig (1995) recently observed that empirical evidence suggests a hierarchy of values for z, depending on the spatial scale of the system being studied. Species-area relationships on continents (or within biogeographic provinces) have values of z around 0.15, those for islands associated with a continent have values between 0.25 and 0.45, while comparisons made among continents or provinces have values around 0.9.

If the concepts regarding geographic range structure are to help unravel the problem of species-area relationships within continents or biogeographic regions, they must produce models that have better explanatory power than models that assume random access of species to local sites. The major difference between a model based on geographic range structure and a random model is that the geographic range structure model assumes that species do not have equal access to local sites, while a random model does not. To examine the consequences of these models for species-area relationships, I conducted numerical experiments on a computer to simulate models that made different assumptions regarding how species access sites. To do this, I constructed a 10×10 grid (units are arbitrary distances) within which I distributed species according to three different sets of rules, as follows:

(1) The first rule I used was a simple random distribution of species among the 100 grid cells. A pool of 500 species was used, and individual species were given equal access to all sites. Each site was assigned as many individuals as it could accept. Assignment of individ-

uals to each site was done as a lottery, so each species had an equal probability of having an individual assigned to each site. Sites were allowed to accept a predetermined number of individuals and all sites had the same number of individuals. Once a site was filled, the number of species at the site were counted. This distributional rule simulates the passive sampling hypothesis.

(2) The second rule I used was one that assumed that each site had a different number of individuals that could be assigned to it. That is, sites were not uniform as in the previous model, so each site accepted a unique number of individuals. Then, as before, individuals were assigned to each site by a lottery. Each species had equal probability of having an individual assigned to any given site. When the site was filled, the number of species with at least one individual at the site was counted. This rule simulates the habitat heterogeneity hypothesis. Recall that this hypothesis includes heterogeneity due both to disturbance and to spatial variation in physical conditions.

(3) The third rule was based on the idea that the geographic range structure paradigm outlined above requires that species do not have equal access to sites. Species vary considerably in the size and shape of their ranges (Maurer 1994; Brown, Taper, and Marquet 1996). I used a relatively simple shape for species' ranges. I used a Gaussian (i.e., bivariate normal) distribution to generate the probability that a species would be found at each of the 100 sites used in previous simulations. The mean of the Gaussian distribution for each species (corresponding to the geographic range center) was chosen from a bivariate uniform distribution. Initial simulations indicated a significant boundary effect existed, so the boundaries for locating range centers were extended beyond the 10×10 grid. Abundances among species approximately follow a truncated lognormal distribution (chap. 6 in this volume; see also Maurer 1994; Brown 1995), so total abundances of each species were generated using such a distribution in the simulations. The mean and variance of this distribution were not changed from simulation to simulation. Since abundance and geographic range size are positively related (Brown 1995), I made the variances of the Gaussian distribution proportional to each species' total abundance. The covariance (or correlation) between latitude and longitude determines the orientation of the geographic range (Maurer 1994). The orientation of each range was determined by choosing a correlation coefficient between -1 and $+1$ from a uniform distribution. Finally, the abundance of each species at each of the 100 sites was determined by multiplying the total abundance of the species by the probability of finding that species at the site. If the abundance of a species at a given site was greater than 1, it was considered to be present on the site; otherwise it was considered absent. This simulation explicitly formalizes the assumption that the size, shape, and orientation of a species' range determines its presence at a given site (see Colwell and Hurtt 1994 for a similar technique applied to latitudinal gradients in species diversity).

Each of the three algorithms discussed above generates a species-area curve. The parameter *z* describes the rate of accumulation of species with area, but it depends on the number of species that occur on each site. To see this, note that if all species occur at all sites, then the value of *z* must be zero, since no new species are found as the area surveyed increases. At the other extreme, if no species occur at any sites, then *z* must be positive infinity, since no amount of increase in area will increase the number of species. In between these extremes, the value of *z* must decrease from infinity to zero, that is, as more species occur at each site, *z*, the rate of accumulation of species with increasing area, must decline. In order to make comparisons among the different simulations, I scaled the number of species found at each site to a proportion of the total species pool (500). The height and possibly the shape of the curve describing the decline of *z* with increasing proportion of the species pool found on each site should be different for each of the three simulation models.

Since each of the three algorithms involves at least some randomness, variation among different runs of each algorithm is to be expected. The randomness for the passive sampling and habitat heterogeneity algorithms comes from the lottery process that assigns species to sites. In the geographic range structure algorithm, the abundance of each species at each site is completely determined by the parameters of the Gaussian distribution. For a given set of parameters, the abundance, and hence the presence, of each species is fixed (i.e., it cannot vary from simulation to simulation). Variation among different runs is generated only by random assignment of the locations of geographic range centers and total abundances for each species. To account for this variation, I repeated all simulations one hundred times.

The results of these simulations indicated that there was a distinct difference between the *z* values generated by the geographic range structure algorithm and the other two (fig. 8.1). The value of *z* declined with increasing proportion of the species pool represented at each site, but it was much lower for the range structure algorithm, falling within the range of continental slopes reported by Rosenzweig (1995).[24] Only when the proportion of the species pool found at each site was high (>.75) did the values generated by the passive sampling and habitat heterogeneity algorithms approach the range of values for continents. That is, the rate of increase in species number with increasing area was low only when most species were found at most sites. The difference between the range structure algorithm and the others is that in the former the number of

24. Note that comparisons between the simulations and Rosenzweig's empirical values are valid, since *z* is independent of units used to measure area (Rosenzweig 1995, 21–22).

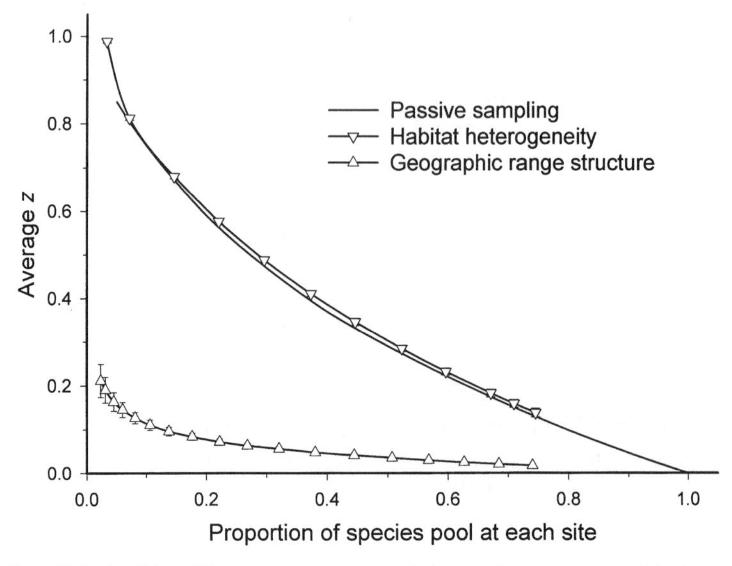

Fig. 8.1 Relationship of the average exponent of the species-area curve with the average proportion of species from the species pool found at a site obtained from simulations incorporating different rules for assigning species to local sites. Note that the geographic range structure rule produces much lower species-area slopes than the other rules when saturation of local sites with species from the species pool is low.

species per site is determined primarily by the range properties of the species, while in the latter the number of species per site is determined by the characteristics of the sites—that is, species are assigned to sites based on how many individuals a particular site can hold. Causation in the range structure algorithm is upper level, that is, the characteristics of sites (species number) are determined by processes that operate at a larger spatial scale than the individual site.[25] In this case, the model incorporating this upper-level form of causation is the model that best explains the data.

The Relationship between Distribution and Abundance

Distribution is often defined as the number of sites (or islands) within a geographic region (or archipelago) occupied by a species. In chapter 6 I discussed the finding that distribution is positively correlated with aver-

25. The properties of a species' geographic range are determined by the set of ecological adaptations possessed by the species. These adaptations are not determined by what happens at any one site. Rather, they are due to differential fitness of variants across many sites. If immigration among sites is sufficient, the most successful of these variants will increase across the geographic range, due either to superior fitness or to a rate of gene flow that overwhelms locally adaptive variants.

age abundance on sites where a species is found: species found on many sites are more abundant on those sites where they occur. This pattern varies a good deal, and often depends on the scale at which comparisons are made. At the geographic scale, however, it appears that it would be reasonable to assume that the relationship is positive. The question of what mechanisms might underlie this pattern arose in chapter 6. Although the question of mechanism is far from answered, I want to show here that the abundance-distribution relationship can also be viewed as a consequence of the geographic range structure of species on a continent.

Using the same methods I used in the previous section to examine the species-area relationship, it is possible to ask if the geographic range structure approach is better at generating abundance-distribution correlations than a passive sampling or sampling with habitat heterogeneity process. I used the same protocol as in the last section to generate a collection of sites with abundances for each of 500 species. In the set of simulations I used to generate species-area relationships, I did not have to obtain an abundance for each species at each site; I simply determined whether it was present or not. To generate a species-abundance relationship, I had to assign an abundance to each species at each site on the 10×10 grid at which it was found. This was straightforward for the geographic range structure rule (rule 3 of the previous section), since the presence of a species depended on whether its estimated abundance from the Gaussian distribution was greater than 1. Rules 1 and 2, however, do not specify an explicit value for abundance. To make these simulations comparable to those based on rule 3, I had to determine a value for abundance for species that occurred at a site. To do this, I chose the total abundance of species at any given site as the average of the total abundances of species at sites obtained in simulations based on rule 3. For rule 1, each site had the same total abundance as other sites. Abundances for each species were assigned by a lottery for "places" at the site. A random number was chosen for each of 500 species, and the species with the highest number in the lottery was assigned one place. This procedure was replicated until the total number of places assigned to all species equaled the predetermined total abundance. Thus, if the total abundance for a particular site was 1,000 individuals, then 1,000 lotteries were conducted. The abundance for each species on any given site theoretically could range from zero to the total number of lotteries conducted for the site, though in practice this never occurred. Generally, the number of species at a site was correlated with the number of individuals at the site. The simulations for rule 2 followed the same procedure, except that the total abundance at each site, instead of being fixed, was randomized.

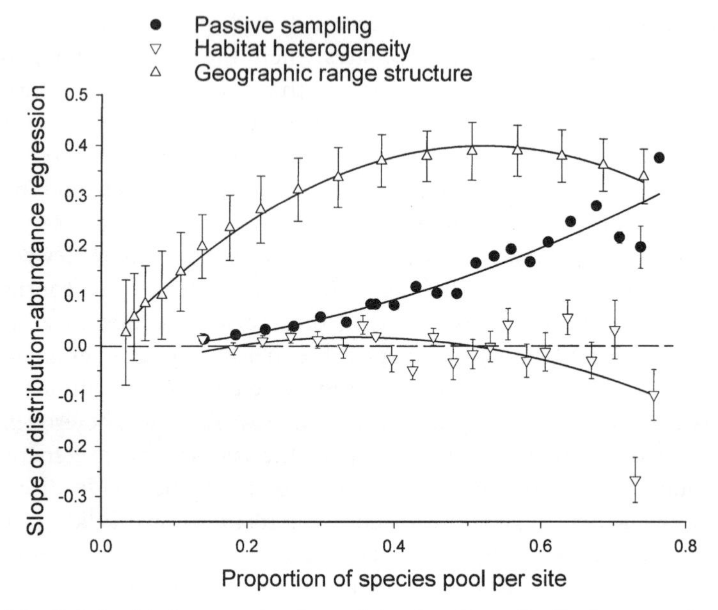

Fig. 8.2 Relationship between the slope of the abundance-distribution regression with the average proportion of species from the species pool found at a site obtained from simulations incorporating different rules for assigning species to local sites.

The results of these simulations indicated that only the geographic range structure rule (rule 3) produced a positive abundance-distribution relationship when the proportion of species at any one site was below .5 (fig. 8.2). When sites were represented by a relatively high proportion of species of the species pool, the passive sampling rule produced positive abundance-distribution correlations, while the habitat heterogeneity rule produced slightly negative correlations (fig. 8.2). The patterns for high levels of saturation of sites with species from the species pool are mildly interesting from a theoretical standpoint, but in real biogeographic systems, it is unlikely that a large fraction of the species in a biota will be found at each local habitat. Hence, the most meaningful results, as is true for the species-area analysis described above, are those for relatively low levels of saturation. At those levels, only the geographic range structure rule produces a correlation between distribution and abundance.

As with the species-area relationship, it appears that the upper-level causal process of the geographic range structure rule is most likely to mimic how the structure of local communities is determined. If this is true, then the reason that abundance is correlated with distribution is that species with ecological adaptations that allow them to have large geographic ranges will also tend to dominate local communities (Brown 1984). Initially this seems reasonable, as long as the resources used by the species with large geographic ranges are roughly equivalent to those used by species with small geographic ranges. In some situations, deviations from this pattern might be expected if widespread species are using entirely different resources than related species that are more narrowly distributed. This might happen, for example, if the widespread species specializes on a resource found in low concentrations across a large geographic region. Then the widespread species might be less common locally than its more narrowly distributed relatives. Such a situation, however, departs from the basic assumptions of the geographic range structure model, which assumes that all species use essentially the same kind of resources, but that some use them more efficiently than others.

Patterns of Nestedness in Species Distributions

The final pattern that I consider as a possible consequence of variation in geographic range structure among ecologically similar, related species is the pattern of "nestedness." This pattern is related to the abundance-distribution correlation examined in the previous section. The basic idea is that sites (or islands) with few species will have subsets of the species that are found on speciose sites (Patterson and Atmar 1986). Those species found only on sites with few species will be the widespread species. Speciose sites will have both widespread and more narrowly distributed species. Of course this pattern is seldom perfect in nature, and therefore distributions are rarely perfectly nested. Indexes of nestedness exist that can be used to describe the degree to which a collection of islands or sites has nested distributions of species (Wright and Reeves 1992; Atmar and Patterson 1993).

For the purposes of this discussion, it is of interest to know whether or not the three rules for constructing local communities described above are able to produce patterns of nestedness. Using each rule as described above, I generated species distribution patterns as before, and calculated Wright and Reeves's (1992) index of nestedness for each assemblage. The results were rather interesting. Neither the passive sampling nor the habitat heterogeneity rules (rules 1 and 2) led to nested patterns of species distribution. For each simulation done using these rules, the value of the Wright-Reeves index was zero, indicating no pattern. Only the geographic range structure rule (rule 3) gave nonzero values for the index (fig. 8.3). When local sites were relatively unsaturated with species from the species pool, the Wright-Reeves index was no lower than about 0.5. As local sites increased in the proportion of the species pool found at the

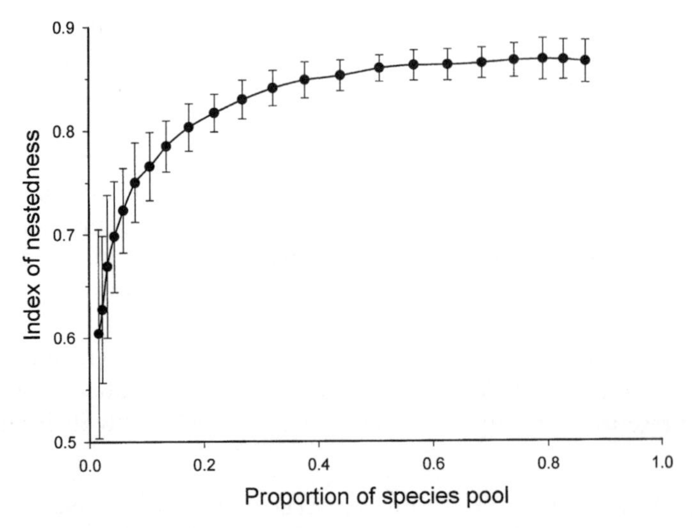

Fig. 8.3 Relationship of Wright and Reeves's (1992) index of nestedness to the average proportion of species from the species pool found at a site obtained from simulations incorporating the geographic range structure rule for assigning species to local sites. Both the passive sampling and habitat heterogeneity rules always gave nestedness index values of zero.

average site, the Wright-Reeves index increased and leveled off at around 0.85.

As with the two patterns discussed above, the model incorporating geographical effects on community structure developed in this chapter offers a simple explanation for a community pattern that has been shown empirically to be a common property of communities within heterogeneous landscapes. Nestedness is a quantitative property of a collection of species assemblages in a landscape, and the results in this section show that it cannot arise from a simple random sampling process. The pattern can arise, however, if species are limited in their geographic distributions.

Geographical Determinants of Community Structure and Synecology

The ideas I have discussed thus far in this chapter have important implications for how community ecology should approach the complexity inherent in its subject material. If a rigorous approach to synecology is to arise from modern community ecology, it must meet two basic requirements. First, it must be based on empirically demonstrable patterns. These patterns must be shown to be derivable from basic premises about the way organisms behave and how they interact with one another and their environment. Furthermore, they cannot be patterns that are simply the consequence of sampling or other purely stochastic processes, although there are undoubtedly stochastic effects that influence community structure. I submit that the patterns discussed thus far in this chapter are candidates for this enterprise. Second, there must be a rigorous theoretical framework that can be used to develop testable hypotheses regarding factors determining the structure of communities. Although such a framework has yet to be fully developed, the results I have discussed above suggest that the framework will need to be explicitly hierarchical, so that causation by upper-level constraint is as much a part of theories as is causation from interactions among system components.

Geographic Assembly Rules

If communities are put together from individualistic responses of species to environmental variations and perturbations, then the only way we can understand them is to understand the ecology of individual species. Since communities are not "wholes" according to the extreme of such a view, then the only thing to study is the parts. This seems to be the basis of much of the disagreement surrounding assembly rules. When it was demonstrated by Connor and Simberloff (1978, 1979) that patterns such as those identified by Diamond (1975) could be generated by simple stochastic processes, many ecologists doubted that there was a set of formal assembly rules that could be used to understand patterns of community structure. The idea that there are assembly rules that determine the composition of a community has been hotly debated since that time. Wiens (1989a) reviewed the details of this debate. It revolves around what kinds of evidence and modeling protocols are necessary to demonstrate the existence of assembly rules.

The approach that I have taken in this chapter suggests that there is indeed a kind of structure that exists in local communities, and that this structure is partially imposed upon the community from outside its boundaries. This is really not a new viewpoint. Ricklefs (1987), for example, stated that ecologists cannot appeal only to local-process theories, such as competition and predation, to understand why a particular set of species exists in a particular place in space and time. He argued that to fully explain community diversity it was necessary to appeal to global processes such as dispersal, speciation, and the history of environmental change. Even MacArthur (1972) devoted a chapter in *Geographical Ecol-*

ogy to the role of history in determining patterns of species diversity. This chapter was an attempt not to dismiss the role of history, but rather to show how ecology might interface with paleontology and biogeography. Themes I have developed in this book, such as the hierarchical structure of ecological systems and the importance of widespread and abundant species, were touched upon by MacArthur (1972).

Given that ecologists have been thinking about larger-scale issues for a long time, is it possible to begin to coalesce their ideas into a framework for understanding, and perhaps eventually predicting, community structure? I think that there are several concepts that might eventually be applied to this problem. Of course there will be other concepts, and the particular usefulness of any one concept may be taxon dependent. But here are a few that I think will improve the ability of community ecologists to make useful predictions regarding community structure.

A community is composed of species drawn from a set of species that exists in the larger geographic region in which the community is located. It is unlikely that any particular subset of species found in any one community is a random draw from the species pool. Different species have different degrees of access to any particular community, depending on where in their geographic range that community is located. Communities near the center of a species' range will be relatively accessible; those near the periphery will be colonized with greater difficulty.[26] Hence, the species list for a community will be partly determined by the subset of the species pool that has geographic range centers close enough to the community to maintain populations either by reproduction or immigration. This presumably can be quantified if we have sufficient information about the species' geographic populations. Thus, geographic access could, in principle, be used to make quantitative predictions regarding local community structure (Cornell and Lawton 1992).

The existence of a site in the center of a species' geographic range is not sufficient to ensure that a population of that species will be able to persist there. Landscapes are always heterogeneous due to a number of complex factors beyond the scope of our discussion (see, e.g., Pickett and White 1985). Hence, even at the center of a species' geographic range, it is likely that there will be a variety of habitats that vary with respect to their suitability for the species. Metapopulation or source-sink population dynamics may be important to the persistence of a species in

26. A colonization event, as defined by MacArthur and Wilson (1967), consists of the establishment of a population that can persist for a short period of time. Exactly how long this must be is not clear. Operationally, persistence should probably be for at least a generation.

the center of its geographic range if its habitat is at all spatially complex. In the center of a species' range, it should be expected that there will be more source populations than at the periphery. That is, habitat suitability of patches in the geographic range center should generally be near the optimum for a species, while those at the periphery should generally be less than optimal. This means that the persistence of a species in a particular community will depend not only on whether that community is in the center of its range, but also on the suitability of the habitat in the particular patch in which the community is located. Habitat suitability is also a factor that is presumably quantifiable, so that relatively precise predictions might be made. A possible complication here is that species often use relatively different kinds of habitat in different parts of their geographic range, and very little is known about how habitat suitability varies geographically (even if we would expect that it should generally parallel abundance; see the discussion in chap. 7).

If a species' ability to persist within a community varies geographically, we might also expect that its success in interspecific interactions will vary. Specifically, we might expect that a population of a species near the center of its geographic range should be more dominant in its community than a population of the same species near its range boundary. In other words, competitive asymmetry should vary geographically with one species dominating in some regions, another dominating in other regions. Quantifying such asymmetry will be difficult. It is here that field experiments might be extremely useful. Mapping the asymmetry of competition among two or more species across geographic space might provide quantitative estimates of gradients in competitive intensity that could be used in conjunction with distributional and habitat suitability data to build sophisticated predictive models of geographic distributions of community structure.

The geography of community structure is a field that is yet in its infancy. A central concept that might guide further studies arises from the idea that species vary in their access to different communities in a geographic region. Some of this variation is due to intrinsic abilities of species to disperse and use resources, some to competitive abilities of species, and some to chance. From the perspective of the local community, this means that any particular community will comprise a unique mix of species. Variation in structure among communities may occur in a predictable fashion if accessibility of species to communities is primarily deterministic. In this case, community composition would not be a random draw from the species pool, since some species will be more likely to be in certain communities than others (due to the location of different communities with respect to the range center). Local communities would be drawn from a species pool that is not "well mixed," that is, where species have unequal probabilities of being found in communities. Contrast this with a well-mixed species pool, where any species has an equal opportunity to be found in any community (not unlike panmixia in population genetics). As in the simulations earlier in this chapter, a well-mixed species pool leads to lottery-type community dynamics. Geographic variation in such communities, rather than showing some type of regular pattern, should be essentially random.

The degree of mixing of a species pool may provide a new way to compare different geographic regions. For example, are species pools in the tropics better mixed than temperate species pools? Are species pools for islands less mixed than those for comparable continental regions? I will not venture answers to these questions here. Additional theory and data are necessary before we can begin to answer them. Suffice it to say that instead of leading to a community ecology devoid of generalizations and interesting questions, questions regarding the degree of mixing in species pools promise new insights and additional conceptual tools to describe the complexity of nature.

Conclusions

In this chapter, I have argued that causation in communities comes from two sources. In the past, theories of community structure have been built primarily from lower-level processes, that is, based solely on the attributes of individuals within populations. Although there is nothing inherently wrong with such an approach, it has its limitations. Those limitations are underscored by the empirical problems described in chapters 3–5. As Salthe (1985) argued, causation can also come from constraints imposed upon a community from larger spatial and temporal scales. It is relatively easy to picture how this might work for environmental causes. Changes in global climate, for example, may shift selection pressures affecting all individuals within a species in the same way, regardless of where they live. Not as easily seen are those constraints that arise from the conservative nature of genetic information. Genetic information can only change by changes in the statistical distribution of genes within a gene pool. This is because genetic information is passed on in a conservative manner from one generation to the next. Due to these genetic constraints, individuals within a species tend to be more similar to one another ecologically than they are to individuals of other species.

The constraints imposed upon a species by the conservative nature of its genetic information play out in space as a characteristic spatial distribution of population abundances. It is these spatial distributions of abundance that constrain what can happen within any single community. Although there may be much stochastic variation among communities, there must be a fairly major component of variation that is attributable to the spatial distributions of abundance of species in the species pool from which each community is drawn. I contend that if community ecologists are successful at integrating current ideas of community structure with an expanded, upper-level approach as called for by MacArthur (1972), Ricklefs (1987), Cornell and Lawton (1992), Brown (1995), Rosenzweig (1995), and many others, a stronger foundation will be established upon which a better understanding of the complexity of ecological communities might be built.

CHAPTER NINE

The Evolution of Species Diversity at the Macroscale

> Therefore, from the species of larger genera tending to vary most & so to give rise to more species, & from their being somewhat less liable to extinction, I believe that the genera now large in any area, are now generally tending to become still larger.
>
> Darwin

The complexity of factors influencing species diversity on local scales makes it necessary to approach the problem of the ultimate factors regulating diversity from a statistical viewpoint. As the scale of interest expands from local to geographic, the details of the many processes influencing local diversity become less important than patterns of change in the environment on geological time scales. Diversity of geographic assemblages of species must ultimately be regulated by differential rates of speciation and extinction (Rosenzweig 1995). Speciation and extinction are processes that result from statistical patterns in ecological processes (Maurer 1989; Maurer and Nott 1998). Extinction is a consistent statistical pattern of negative population growth. Speciation involves a statistical trend of successful establishment of viable populations by an emerging species. Most of the barriers that isolate species from one another can be considered ecological (Brown 1995).

The processes that generate diversity have been discussed in the literature by researchers with different perspectives. On the one hand, some paleontologists have viewed the mechanisms that regulate diversity to be simply the aggregated consequences of microevolutionary processes (Hoffman 1989). This reductionist approach to macroevolution denies any unique contribution to the regulation of diversity to processes occurring at larger scales: such processes are considered epiphenomenal. On the other hand, some have argued that the processes that shape macroevolution are decoupled from microevolutionary or ecological processes (Gould and Eldredge 1977; Stanley 1982, 1990). This argument has been taken so far as to suggest that although with hindsight we might be able to identify why groups of species evolved the way they did, the real world is so complex that were it possible to return to a previous point in the evolutionary history of a clade and restart its evolution, the subsequent history of the clade would be completely different from its observed history (Gould 1989). These two positions should be considered as extremes; there are many perspectives that incorporate intermediate views (Salthe 1985; Eldredge 1989; Maurer 1989; Maurer, Brown, and Rusler 1992).

The approach to communities that I have taken thus far lends itself to an examination of the middle ground between the microevolutionary extrapolationist and the macroevolutionary elitist positions on the question of the evolution of diversity. I will begin this examination with a rereading of some of Darwin's views of evolution. It is evident that Darwin thought of evolution as a complex process involving at least two different interacting processes occurring on different scales. Next, I develop an analytical approach to the evolution of diversity and discuss some consequences of that approach. Finally, I will try to apply some of the insights obtained from this approach in an analysis of data on Neogene planktonic foraminifera (Stanley, Wetmore, and Kennett 1988).

Darwin's Two Evolutionary Mechanisms

In this section I will examine Darwin's ideas about how evolution operates. I will argue that one way to interpret Darwin's evolutionary theory is that he saw natural selection as determining to a large degree the rates of extinction within groups of related organisms. But Darwin's concept of evolution relies heavily on the individuality of species, because, in his view, species compete with one another, and eventually the most widespread species and higher taxa replace the more narrowly distributed taxa.

Natural Selection

Most treatments of Darwin's theory of evolution have emphasized his ideas regarding what has come to be known as microevolution. These ideas formed Darwin's mechanism of evolution that he called *natural selection.* One of the more succinct descriptions of this mechanism is given in the first chapter of *The Origin of Species:* "As many more individuals of each species are born than can possibly survive; and as, consequently, there is a frequently recurring struggle for existence, it follows that any being, if it vary however slightly in any manner profitable to itself, under the complex and sometimes varying conditions of life, will have a better chance of surviving, and thus be *naturally selected.* From the strong principle of inheritance, any selected variety will tend to propagate its new and modified form" (Darwin 1859, 5; italics in original). This idea of differential proliferation of heritable forms of variation has formed the backbone of the modern science of evolution. Advances in the understanding of the mechanisms of inheritance have added a great deal to our understanding of evolution since Darwin's time. Numerous mechanisms other than natural selection are known to produce changes in the distribution of characteristics in populations, yet the theory of evolution by natural selection has stood the test of time. Natural selection is a measurable phenomenon (Endler 1986), not just a theoretical abstraction. It is clear that natural selection has contributed to the evolution of many characteristics, although controversy still exists regarding its ultimate role in evolution.

The two different extreme positions regarding macroevolution alluded to above have very different views regarding the importance of natural selection in evolution. The microevolutionary extrapolationist position sees all of evolution as having resulted from the operation of natural selection over geological time scales. Under this view, differences among related taxa are due to the operation of natural selection on different populations of a common ancestor. Consequently, diversity can be considered an epiphenomenal consequence of the operation of natural selection over long periods of time. The macroevolutionary elitist position takes the opposite view. Natural selection, according to this view, is a transient process that occurs uniquely within a few populations to adjust the distribution of characteristics in the population to local conditions. Occasionally, natural selection may be involved in the change of some characteristics of relevance to the survival of a species during speciation, but this need not contribute much to the history of evolution: other mechanisms can operate just as efficiently to cause species to diverge from one another.

Differential Proliferation of Species

Darwin's theory of evolution was more complex than the suggestion that the operation of natural selection could lead to adaptive changes in populations of organisms. He also had in mind a second mechanism based on the effects of natural selection on the divergence of varieties of the same species and the differential proliferation of widespread, dominant clades as compared to narrowly distributed clades.

In the second chapter of *The Origin of Species,* Darwin set out his observations regarding the differences in variability between widespread and narrowly distributed genera:

> in any limited country, the species which are most common, that is abound most in individuals, and the species which are most widely diffused within their own country (and this is a different consideration from wide range, and to a certain extent from commonness), often give rise to varieties sufficiently well marked to have been recorded in botanical works. Hence, it is the most flourishing, or, as they may be called, the dominant species,—those which range widely over the world, are the most diffused in their own country, and are the most numerous in individuals,—which oftenest produce well-marked varieties, or, as I consider them, incipient species. (1859, 53–54)

In the above passage, Darwin was reporting the results of tabulations that he did of several well-studied floras. He did not give the details of which floras he studied in either the first edition, from which the above quote is taken, or the sixth edition; but in both cases he promised a more detailed treatment in a later publication. Although he never published these calculations, the manuscript of Darwin's notes on these calculations has been published (Stauffer 1975). Darwin's generalization regarding the positive relationship between distribution and commonness has been clearly borne out by a number of detailed studies on plants and animals (see chap. 6). What is particularly interesting, however, is his suggestion that those species which are widespread are also more variable. Recent genetic evidence suggests that there is a tendency for widespread species to be more variable than closely related species of narrow geographic distribution (Bowers, Baker, and Smith 1973); however, there is relatively little data to indicate how general this pattern is. Certainly, here is a point upon which molecular biology and population biology could be jointly applied to the benefit of both fields.

The mechanism that Darwin proposed for this difference in variability between common and rare species incorporated natural selection: "for, as varieties, in order to become in any degree permanent, necessarily have to struggle with the other inhabitants of the country, the species which are already dominant will be the most likely to yield offspring which, though in some slight degree modified, will still inherit those advantages that enabled their parent to become dominant over their compatriots" (1859, 54). In other words, Darwin thought that widespread species were capable of generating more varieties because they were able to dominate less-widespread species ecologically throughout their geographic range. Natural selection could then allow these varieties to diversify as they adjusted to local conditions.

The difference in variability between common and rare species had important consequences for Darwin's view of evolution. Natural selection was responsible for adjusting organisms to the varying conditions of life, but the differences between common and rare genera determined which kinds of organisms ultimately dominated a biota. A species that arose from speciose and widespread genera, according to Darwin, would tend to inherit the characteristics that made the genus widespread and dominant in the first place. Such a species would have an advantage over species from less-widespread genera. In his chapter on natural selection, Darwin recapitulated his argument regarding variation: "We have seen that in each country it is the species of the larger genera which oftenest present varieties or incipient species. This, indeed, might have been expected; for as natural selection acts through one form having some advantage over other forms in the struggle for existence, it will chiefly act on those which already have some advantage; and the largeness of any group shows that its species have inherited from a common ancestor some advantage in common" (1859, 125). Note here that Darwin again recognized the importance of species of successful groups inheriting characteristics that give them an advantage when compared to other species. We shall see this concept come out very clearly in the theoretical treatment later in the chapter. Darwin then concluded:

> Hence the struggle for the production of new and modified descendants will mainly lie between the larger groups, which are all trying to increase in number. One large group will slowly conquer another large group, reduce its numbers, and thus lessen its chance of further variation and improvement. Within the same large group, the later and more highly perfected sub-groups, from branching out and seizing on the many new places in the polity of Nature, will constantly tend to supplant and destroy the earlier and less improved sub-groups. Small and broken groups and sub-groups will finally tend to disappear. Looking into the future, we can predict that the groups which are now triumphant, and which as yet have suffered least extinction, will for a long period continue to increase. (125–26)

Here is the clearest statement of Darwin's second evolutionary mechanism. Differential ecological success of species within Darwin's groups, which we would now call clades, leads to differential rates of extinction between groups. The consequences of natural selection are in part predictable. Although we cannot predict when an ecological innovation will occur within a clade, according to Darwin, those innovations that confer upon the species that possess them an advantage will cause the diversity of the clade to increase over time, while other clades, with less-efficient ecological adaptations, will be less successful and eventually become extinct. Any dominant group will continue to increase in diversity until ecological conditions change to favor another clade or a superior ecological innovation arises in a different clade. The key to predicting the direction of the evolution of diversity, according to Darwin, is the "fit" of the adaptations of the species to current ecological conditions.

It should be noted here that Darwin's second evolutionary mechanism was not simply an extrapolation of natural selection operating over long periods of time. Browne (1980) argued that Darwin needed more than natural selection to explain the patterns of commonness and rarity that he documented. She pointed out that Darwin originally tried to explain the tendency of species to diverge as a consequence of natural selection, but added sections on the diversity of widespread genera and on the principle of divergence to his unpublished manuscript after he realized that larger genera must be more successful than smaller genera. When he abstracted the *Origin* from this larger manuscript, he included the passages quoted above, which came from these sections. Conceptually, Darwin considered the different rates of proliferation of widespread and rare taxa to be a distinct mechanism of evolutionary change that incorporated natural selection.

Notice that Darwin's idealization of the evolution of diversity includes the metaphor that species are struggling against one another, and that those that are successful in the struggle have been better competitors for limited resources. This concept incorporates much of what has come to be known as competition theory in the ecological literature. Although interspecific competition, as envisioned by MacArthur (1972), for example, is clearly a mechanism by which the differential success of species can lead to changes in diversity, it is fruitful to consider Darwin's "struggle for existence" in a larger context. Each species might be conceived of as struggling to survive in a constantly changing environment, of which other species that use similar resources are a part. The success of the struggle depends not only on adaptations that allow species to partition resources, but on many other kinds of adaptations that influence the life histories of individual organisms.

The Evolution of Diversity: The Ecology of Speciation and Extinction

Darwin had no clear idea of how speciation occurred. He simply assumed that as species were slowly changed by natural selection they would eventually become different enough to form well-defined species. Although modern findings regarding the mechanisms of speciation might require us to modify our perception of how species form, Darwin's argument regarding which species persist and have opportunities to further speciate is still valid. In the following section I consider some of the implications of Darwin's second evolutionary mechanism.

Given Darwin's conclusions regarding the differential proliferation of different taxa based on their ecological success, it is fruitful to consider the ecological contexts in which speciation and extinction occur. A number of people have attempted to do this, and I have reviewed some of these attempts elsewhere (Maurer 1989, 134–36). Here I consider the subject from a somewhat different perspective, following the line of reasoning that Vrba (1980) and Eldredge (1989) have developed.

Recall that in chapter 7 I argued that species that are widespread and abundant are more resistant to extinction than more narrowly distributed ones. The evolution of diversity is directly tied into the ecological success of species as indicated by their geographic range structure. The basic idea is that because the environment used by a species has a limited amount of resources that the species can use, the ecological limitations that result, coupled with the distribution of ecological adaptations within the species, influence its likelihood of persistence and its ability to give rise to new species. Species with the ability to use a wide variety of resources can develop relatively dense populations locally and will have large geographic ranges. Consequently, species with broad ecological tolerances will not be as prone to extinction as species with narrow requirements. Furthermore, species with wide niches may also have a higher likelihood of having small, isolated populations survive to become new species. Hence, the diversity of clades composed of species with wide niches should be greater than that of clades in which species have narrow niches (Maurer and Nott 1998).

This interpretation differs in some respects from Eldredge's (1989) treatment of the subject. Eldredge used the term *stenotope* to refer to what I call species with narrow niches, and *eurytope* to refer to species that have broad niches. Eldredge considers Vrba's (1980) study of two sister clades of fossil and recent African antelopes in examining the relationship between ecological breadth and the likelihood of speciation and extinction. Vrba argued that the two sister clades differed in diversity as a consequence of different ecological breadths. According to Vrba, during the time since their divergence from each other in the Miocene, the impalas have been geographically widespread, but have had low diversity, while the wildebeests have been more narrowly distributed, but more diverse. Vrba argued that the wildebeests were more diverse because their more specialized ecological adaptations made it more likely for them to become isolated. She went further to postulate that because the wildebeests had a narrow ecological spectrum, they also might have had a tendency to be more selective about the mates they chose. This would lead to an increased tendency for stenotopes, such as the wildebeests, to evolve reproductive barriers through differential mate selection, and hence have a higher speciation rate.

Eldredge (1989) argued that stenotopes may actually have higher speciation rates not because they have a tendency to be more selective of mates, but because in being narrow niched, they are more likely to coexist with closely related, stenotopic species that are already present in the environment. Thus, the likelihood of a newly evolved species persisting is higher if it is more specialized.

Both Vrba's and Eldredge's arguments hinge on the observation that the wildebeests and their close relatives are more stenotopic than the impala. However, it is not entirely clear that this is the case. Censuses of bovid species in eastern and southern Africa reported by Greenacre and Vrba (1984) suggest that both the impala (genus *Aepyceros,* tribe Aepycerotini) and the wildebeest (genus *Connochaetes,* tribe Alcelaphini) are nearly equally dense, and both are apparently geographically widespread (table 9.1). Although they do not report densities for individual species, it is likely that most of the individual wildebeests counted by Greenacre and Vrba belong to a single species (*C. taurinus*). *C. taurinus* is a polytypic species (Ansell 1971) with one widespread form surrounded geographically by a number of more narrowly distributed forms. The second species in the genus, *C. gnou,* is monotypic and also found geographically peripheral to the range of the most common subspecies of *C. taurinus.* Hence, rather than a clade of narrowly distributed stenotopes, as Eldredge and Vrba thought, the wildebeests actually appear to be a group of narrowly distributed forms clustered geographically around a widespread and common species. Habitat use is apparently not a good indicator of ecological success in these two clades.

Clearly there is much more to be learned about speciation in widespread and narrowly distributed forms, but Eldredge (1989) argued that extinction should be higher in narrowly distributed stenotopes than in widespread eurytopes. As we shall see in a moment, what is important in determining the diversity of a clade is the difference between speciation and extinction rates, not the size of these rates per se. The differences I discussed regarding variation among species with large and small geographic ranges lead to the prediction that eurytopes, although they may have relatively low speciation rates, will also have low extinction rates. Clades of high diversity will be descended from eurytopes because the difference between speciation and extinction rates will be greater for eurytopes than for stenotopes. This argument recalls Darwin's original ideas regarding the success of widespread taxa discussed above.

Table 9.1 Characteristics of some African bovid genera and their distribution on 16 African game reserves.

Tribe	Genus	Number of Species	Average Density (no./km^2)	Number of Reserves Occupied	Average Biomass (kg/km^2)	Body Mass (kg)
Alcelaphini	*Damaliscus*	3	0.34	3	38.1	112
	Connochaetes	2	4.16	12	802.9	193
	Alcelaphus	2	1.78	6	283.0	159
Aepycerotini	*Aepyceros*	1	6.17	9	388.7	63
Antilopini	*Antidorcas*	1	0.59	2	20.3	34
	Gazella	10	7.11	3	344.9	48
Hippotragini	*Oryx*	3	0.33	4	50.5	155
	Hippotragus	3	0.14	9	30.9	225

Toward a Theory of Macroevolutionary Dynamics

The discussion to this point has suggested that the dynamics of diversity in an evolving clade of species involves both cladogenetic processes, such as speciation and extinction, and population processes, such as natural selection. We can extend Darwin's insight regarding the relationship between natural selection and the differential proliferation of species by considering models that tie population dynamics into changes in the number of species in an evolving clade over time. Our interest, as in previous chapters, will be to obtain relationships between processes occurring on different scales. For the present discussion, this means that we want to obtain a relationship between the dynamics of speciation and extinction on the one hand and the dynamics of population change on the other. As we will see, this relationship is a statistical one that reflects constraints that each level imposes on the other.

We are interested, for the present discussion, in the patterns of accumulation of diversity over time rather than the accumulation of phenotypic change. Hence, we are interested in taxonomic macroevolution rather than transformational macroevolution (Eldredge 1979). The arguments developed above might be applied to either phenomenon, and I suspect that the general conclusion regarding the role of population-level processes would be similar in each case. However, I will deal in this chapter with the accumulation of diversity rather than the accumulation of phenotypic change, because the connection between different levels is more apparent in the model describing the evolution of diversity.

A Statistical Model of the Evolution of Diversity

We begin by assuming that the number of species in a given clade is a function of the total population size of each species. In other words, we assume that cladogenetic processes are ultimately ecological in nature. Eldredge and Gould pointed out that simple extrapolations of the results of anagenetic processes to determine macroevolutionary patterns fail to recognize that speciation is an "ecological and geographic process" (1972, 96). That is, regardless of the genetic mechanisms that lead to the isolation of a gene pool, the incipient species must successfully establish itself ecologically over a sufficiently large geographical area in order to be recognizable as a new species. If it cannot establish itself ecologically, then it will become extinct before it can be considered a distinct species. Although such a process may extend over a relatively long period in ecological time, it can be viewed as a nearly instantaneous event in evolutionary time. In a similar way, extinction events may occur relatively rapidly in evolutionary time. Extinction is an ecological process because it is the end point of sustained negative population changes over ecological time. In evolutionary time, such changes may appear to be instantaneous. Because speciation and extinction will be instantaneous in evolutionary time, likewise changes in species number in a clade might be viewed as a relatively continuous process. Changes in taxonomic diversity over time have been successfully modeled as a continuous process in the past (Maurer 1989; Rosenzweig 1975; Sepkoski 1978; Walker 1985).

The number of species in a clade at a given point in time, $S(t)$, can be written as the sum of indicator functions (see, e.g., Lasota and Mackey 1994, 5), that is,

$$S(t) = \sum_{i=1}^{S_{max}} 1_i^t(N_1, N_2, \ldots, N_{S_{max}}), \qquad \textbf{(9.1A)}$$

where S_{max} is the largest number of species that could be found in the clade and

$$1_i^t(N_1, N_2, \ldots, N_{S_{max}}) = \begin{cases} 1 & \text{if } N_i > 0 \text{ at time } t, \\ 0 & \text{otherwise.} \end{cases} \qquad \textbf{(9.1B)}$$

There are some important limitations on the indicator functions given in equation (9.1B) that must be imposed in order to make them sensible in terms of species dynamics. First, we assume that the indicator functions are random variables, that is, they represent a statistical distribution of possible outcomes based on the current species abundances. Second, there will only be $S(t)$ species at time t that have nonzero population sizes, and, to make sense physically, the functions 1_i^t must be defined so that they are determined only by the species present. That is, a species that has not yet come into existence through speciation cannot affect its own, or any other species', likelihood of being included in the future. Another way of interpreting equation (9.1B) is to say that the presence of species i at time t depends on the abundances of all of the other species present at the same time and on the ecological characteristics that it possesses. Note that at a different time, say $t + \Delta t$, species composition might have changed due to extinctions and speciation events. Thus, if $1_i^t = 1$ and $1_i^{t+\Delta t} = 0$, then species i became extinct during time interval Δt. Likewise, if $1_i^t = 0$ and $1_i^{t+\Delta t} = 1$, then species i originated during Δt. Notice also that the indicators are functions of time, that is, they may change from time to time as ecological and physical conditions in the environment change. The formality of letting the indicator functions be functions of S_{max} allows us to add and delete species as they speciate and become extinct.

It is difficult to write a dynamical description of changes in species number over time because $S(t)$ is a discrete number and the indicator functions are also discrete. This can be circumvented by considering the statistical behavior of $S(t)$. We examine the behavior of the expected value, or average, of $S(t)$, where $S = \mathrm{E}[S(t)]$. The expected value of an indicator random variable is a probability, that is, $\mathrm{E}[1_i^t] = p_i$, where p_i is the probability of species i being present in the clade. These probabilities should explicitly be considered to be a function of population sizes (the N_i's). We will not write them as such to simplify notation. Thus, the average number of species, S, is

$$S = \sum_{i=1}^{S_{\max}} p_i. \tag{9.2A}$$

The dynamics of average species number in the clade will be

$$\frac{dS}{dt} = \sum_{i=1}^{S_{\max}} \frac{dp_i}{dt}. \tag{9.2B}$$

Now recall that the p_i's are functions of the population dynamics of each species, so for each probability we can write

$$\frac{dp_i}{dt} = \sum_{k=1}^{S_{\max}} \frac{\partial p_i}{\partial N_k} \frac{dN_{ik}}{dt},$$

so equation (9.2B) can be written as

$$\frac{dS}{dt} = \sum_{i=1}^{S_{\max}} \sum_{k=1}^{S_{\max}} \frac{\partial p_i}{\partial N_k} \frac{dN_k}{dt}. \tag{9.3}$$

Equation (9.3) explicitly connects the dynamics of species diversity of the clade with the total population sizes of each species in the clade. The terms $\partial p_i/\partial N_k$ represent the effects of population dynamics of species on the probability that a species will be found in the clade. These effects are twofold: first, the population size of a species will be related to the likelihood of its own extinction, and in some speciation models may influence the likelihood that it gives rise to new species; second, population size of a species may influence the likelihood that related species will become extinct via competition (Maurer 1985a) and may also affect speciation if competitive speciation occurs in nature (Pimm 1979; Rosenzweig 1978). Thus, equation (9.3) describes the ecological processes that are involved in changes in the number of species in a clade over evolutionary time. Note that we are concerned here with *total* population sizes of species across their entire geographic range. These, in principle, are very different from the dynamics that occur within individual populations or metapopulations within a species' geographic range. Since most species in the fossil record persist for hundreds of thousands to millions of years, the time scale of population dynamics is much longer than the time scale for dynamics of smaller population units.

Since it is not possible to measure population abundances of species in the fossil record over evolutionary time, it is necessary to describe changes in species dynamics in terms of per species rates of change. That is, let $\rho(S)$ be the per species rate of change in species number. Then dS/dt can be described with data from the fossil record as species number times the per species rate of change, that is,

$$\frac{dS}{dt} = S\,\rho(S). \tag{9.4}$$

The function ρ is usually expressed in terms of a per species speciation and extinction rate, that is,

$$\rho(S) = \zeta(S) - \xi(S), \tag{9.5}$$

where $\zeta(S)$ is the per species rate of speciation and $\xi(S)$ is the per species rate of extinction (MacArthur 1969; Maurer 1989; Rosenzweig 1975; Sepkoski 1978; Stanley 1979; Stenseth and Maynard Smith 1984; Stenseth 1985; Maurer and Nott 1998). The simplest formulations for the functions ζ and ξ make them linear functions of S, with ζ decreasing and ξ increasing with increasing species number. This gives a logistic model of species dynamics (Maurer 1989; Sepkoski 1978). Recall that in equation (9.3) we defined dS/dt in terms of the population processes of each species in the clade. If equation (9.4) represents a description of species dynamics in terms of per species rates of change, then there is an equivalence between that description and the description obtained by considering the ecological mechanisms that generate those dynamics. That is, since we have two expressions for dS/dt, one that corresponds to mechanisms that we cannot measure (eq. [9.3]) and one that corresponds to patterns that we can measure (eq. [9.4]), we can obtain an expression that relates the measurable pattern to the mechanisms that generate the pattern by setting equations (9.3) and (9.4) equal to one another. Doing this and rearranging the result gives

$$\rho(S) = \frac{\sum_{i=1}^{S_{\max}} \sum_{k=1}^{S_{\max}} \frac{\partial p_i}{\partial N_k} \frac{dN_k}{dt}}{S}. \tag{9.6}$$

That is, the measurable rate of taxonomic change (ρ) is an average of the changes in species number due to ecological processes. Even though we cannot measure the actual contribution to species dynamics due to any particular species, the rates of change that we can measure correspond to a statistical description of those ecological effects averaged across all species.

The relationship between the measurable patterns of species dynamics and the actual processes governing them has as a consequence an interesting property. Since all clades begin with a single species, when $S = 1$, equation (9.6) becomes

$$\rho(1) = \frac{\partial P_1}{\partial N_1}\frac{dN_1}{dt}, \qquad \textbf{(9.7A)}$$

since we assume that all of the other terms in the sum will be zero. This can also be written

$$\zeta(1) - \xi(1) = \frac{\partial P_1}{\partial N_1}\frac{dN_1}{dt}. \qquad \textbf{(9.7B)}$$

This means that the rate of diversification of the clade, as described by the function $\rho(S)$, is directly related to the ecological characteristics of the founding species of the clade. These ecological characteristics determine the magnitude and form of the speciation and extinction rates that will govern the dynamics of species number in the clade.

Note that in equation (9.7), since $S = 1$, only the parameters determining the function ρ (or alternatively ζ and ξ) are included in the terms on the left-hand side. There are two types of parameters for the unknown function(s) ρ (or ζ and ξ). The first type of parameter contributes to the magnitude of ρ, and the second, to its shape. Since we will be comparing sister species (that is, two species derived from the same common ancestor species), we assume that the parameters determining the shape of ρ are similar among species and, thus, that species differ primarily due to the magnitude of ρ. In Maurer 1989 I suggested that $\zeta(1)$ and $\xi(1)$ could be interpreted as probabilities of speciation and extinction, respectively, of the founding species of a clade. That is, $\rho(1)$ represents the magnitude of the difference between the initial speciation and extinction probabilities of the founding species of a clade. Equation (9.7) implies, therefore, that the probabilities of speciation and extinction of the founding species of a clade are determined by the ecological dynamics of that species (i.e., by dN_1/dt). If two sister species differ only in the magnitude of their per species speciation and extinction rates, then it follows that the clade descended from the species with the larger initial difference between specia-

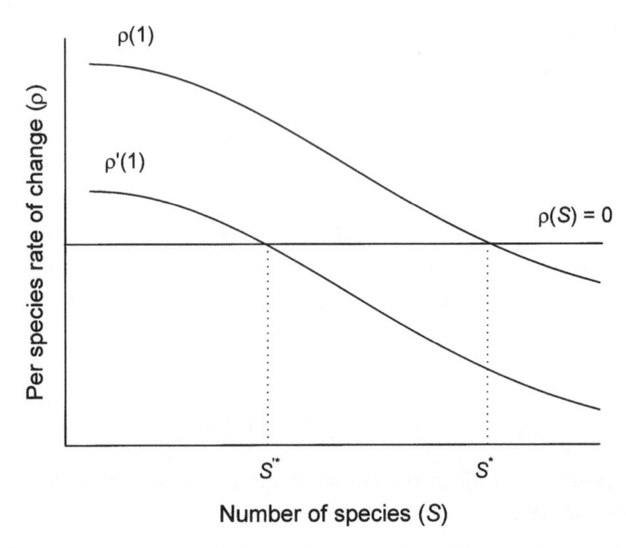

Fig. 9.1 Two per species rate of change functions that differ only in magnitude. Since they are both the same shape, the clade governed by $\rho(S)$ will have a higher equilibrium number of species (S^*) than the clade governed by $\rho'(S)$. Furthermore, for any number of species in the clade, the clade governed by $\rho(S)$ will always have the highest rate of diversification.

tion and extinction probabilities will tend to have more species in it than the clade derived from the other species (fig. 9.1). Over the fossil history of a clade, then, more-diverse clades—that is, clades where on average more species are present at a given time—should be derived from species that had higher per species diversification rates than the ancestors of less-diverse clades.

The prediction stated in the last paragraph is a restatement of one of the major predictions Darwin made in describing the logical consequences of his evolutionary model. Recall that Darwin predicted that the struggle for existence should lie between the larger genera, that is, between genera that had relatively large numbers of species. In the model, I emphasized that the same prediction is made by the explicit connection between the ecological dynamics of the founding species of a clade and the speciation and extinction rates that govern the clade's subsequent diversification, as given in equation (9.7). With appropriate data on patterns of diversification of a clade in the fossil record, it should be possible to quantitatively test this prediction.

Stanley (1979, 1990) has shown that there is a positive relationship between rates of speciation and extinction in a number of taxa. What is important to understand here is that the model developed above suggests that it is the initial difference between speciation and extinction that is important, not the magnitudes of these rates. For example, it is possible for a clade with a low speciation rate to be more diverse than a clade with a high speciation rate if the difference between speciation and extinction rates is greater in the first clade than in the second. It should also be noted that the difference between rates should be greatest when there is only a single species in the clade. In other words, when a clade originates from a species with high speciation potential relative to its likelihood of extinction, it will subsequently have relatively high diversity. As argued above, this is most likely for species that are widespread and abundant.

Testing the Model

We can test a key prediction of the model developed in the last section, namely, the prediction that diverse clades should originate from species with high speciation potential and low extinction likelihood when compared to less-diverse clades. There are, however, several problems with making such a test. First, it is difficult to determine the per species speciation and extinction rates of individual species. As argued above, these rates should be coupled with ecological and biogeographic properties of species, such as population density, geographic range size, and ecological specialization; but the relationship between these variables and speciation and extinction rates is not straightforward and depends on the specifics of the autecology of each species. Certainly, we have nothing close to a theoretical understanding of the quantitative relationship between these ecological variables and speciation and extinction rates. Second, a strong test of the prediction requires data from clades of species for which there is an adequate fossil record. The fossil record must be resolvable at the species level. Third, it is probably not valid to compare clades with distant genealogical relationships (Brooks and McLennan 1991). Comparisons of diversity and rates of speciation and extinction are most valid if made between sister clades (that is, clades which are each other's closest relatives). The most obvious reason for this is that sister clades will share many ecological features with one another by virtue of their common descent. Those aspects of their ecologies that are different are likely to be related to differences in the rates of diversification. Such comparisons of ecological differences cannot be made among distantly related taxa.

The best way to deal with the first difficulty is to realize that per species rates of speciation and extinction are really properties of the entire clade. Thus, data on these rates should be obtained from the entire clade, and then compared or extrapolated back to a single species. In fact, if a particular model seems appropriate for a given set of data, statistical estimates of the parameters of that model can be used to test the predictions of the general model developed above. That is, if specific forms for the functions $\zeta(S)$ and $\xi(S)$ are applicable to a particular data set, then one can compare estimates of $\zeta(1) - \xi(1)$ between clades that differ in diversity.

The second two difficulties can be dealt with by selection of an appropriate data set. Unfortunately, there are few data sets available that are complete enough to allow reliable estimation of $\zeta(S)$ and $\xi(S)$ among sister clades. One exceptional data set was compiled by Stanley, Wetmore, and Kennett (1988) from Kennett and Srinvasan's (1983) monograph on Neogene planktonic foraminifera. Kennett and Srinvasan gave the stratigraphic ranges for 44 species of globigerinid and 50 globorotaliid foraminifera (fig. 9.2). Foraminifera are unicellular organisms that live in freshwater and marine planktonic environments. From what is known of their ecological characteristics, globorotaliids are found deeper in the water column in marine environments than globigerinids, and appear to be relatively specialized herbivores in regions of upwelling as opposed to the more generalized, carnivorous globigerinids (Stanley, Wetmore, and Kennett 1988). The globigerinids generally maintained higher levels of diversity than the globorotaliids throughout most of the Neogene.

Stanley, Wetmore, and Kennett (1988) constructed survivorship curves for each group of foraminifers and found that the globigerinids had lower rates of extinction than the globorotaliids. They also cited evidence that speciation was higher in the globorotaliids. The lower rate of speciation and extinction of the ecologically generalized globigerinids is associated with their higher diversity. This does not agree with Vrba's (1980) and Eldredge's (1989) conclusions. A closer look at the data for the foraminifera using the model developed above helps to demonstrate how estimation of the difference between the per species speciation and extinction rates when only a single species is in the clade clarifies why the more ecologically general clade is more diverse. To do this, it is necessary to obtain valid estimates of per species rates as a function of number of species in the clade, then use these to estimate the rate functions $\zeta(S)$ and $\xi(S)$.

Most often, estimates of per species rates of speciation and extinction are calculated by dividing the number of speciations or extinctions during a time interval by the number of species present during the time interval (e.g., Stanley 1979; Stanley, Wetmore, and Kennett 1988). This is an incorrect procedure in a dynamic system undergoing continuous changes. Suppose that there are S_t species present at time t and $S_{t+\delta t}$ species present at time $t + \delta t$. The per species rate of change is $(1/S)(dS/$

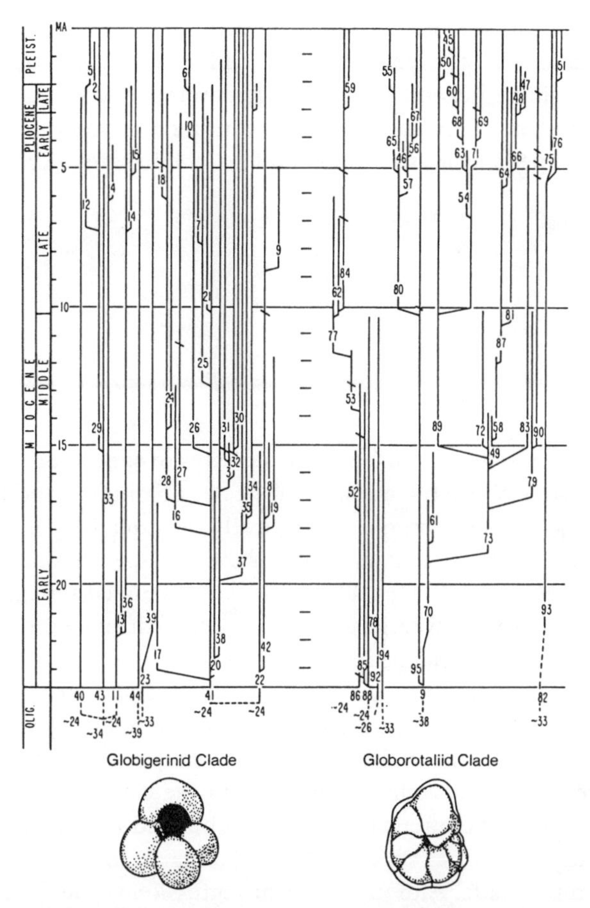

Fig. 9.2 Temporal distribution of species in two clades of Neogene planktonic foraminifera. Speciation and extinction rates were generally more rapid in the globorotaliid clade, but the globigerinid clade generally had more species extant at the same time. From Stanley 1990.

$dt) = d \ln S/dt$. During the time interval t to $t+\delta t$, the average per species rate of change is given by

$$r = \frac{\ln S_{t+\delta t} - \ln S_t}{\delta t}, \tag{9.8}$$

where r is the average per species rate. The quantity r is an estimate of $\zeta(S) - \xi(S)$ during the time interval.

The data for the two clades of foraminifera studied by Stanley, Wetmore, and Kennett (1988) were divided into 24 one-million-year intervals, and the number of species at the beginning of each interval was counted (notice that the length of the time interval δt is 1 million years, so the rates will be per million years). The number of speciation events and the number of extinction events during the interval were also counted. The per species speciation and extinction rates were calculated as

$$r_s = \ln\left(1 + \frac{N_s}{S}\right),$$
$$r_e = \ln\left(1 - \frac{N_e}{S}\right), \tag{9.9}$$

where r_s and r_e are the average per species rates of speciation and extinction, and N_s and N_e are the number of species originating and becoming extinct during the time interval, respectively (Maurer 1989). The difference between r_s and r_e is an estimate of the per species rate of change during the time interval. This difference was calculated for each interval and plotted against the average number of species during each time interval for both foraminiferal clades (fig. 9.2).

For both the globigerinid and globorotaliid clades, there appears to be a linear relationship between the per species rate of change and species number (fig. 9.3). This makes it relatively simple to estimate the per species rate when only a single species is in the clade. This estimate is simply the sum of the intercept and slope obtained from a simple linear regression of average per species rate of change against average number of species in the clade. Although a linear regression model seems appropriate here, a nonlinear model could also have been used (see, e.g., Maurer 1989). The difference in the per species rates of change with one species present is significantly higher for the globigerinids than it is for the globorotaliids ($t = 3.95$, $df = 46$, $P < 0.001$, one-tailed test). The ecologically general globigerinids also maintained a higher number of species during the Neogene than the more specialized globorotaliids (Stanley, Wetmore, and Kennett 1988). Hence, the data for the Neogene foraminifera support Darwin's model of diversification rather than the model implicit in the assumptions of Vrba and Eldredge.

Stanley (1986) analyzed another data set that provides additional support for Darwin's model. He compared the rates of extinction between siphonate and nonsiphonate bivalve mollusks in two Neogene fossil assemblages using a statistic called Lyellian percentages. A Lyellian percentage is that percent of a given fossil fauna that has survived to the present (see also Stanley 1979). Stanley (1986) concluded that popula-

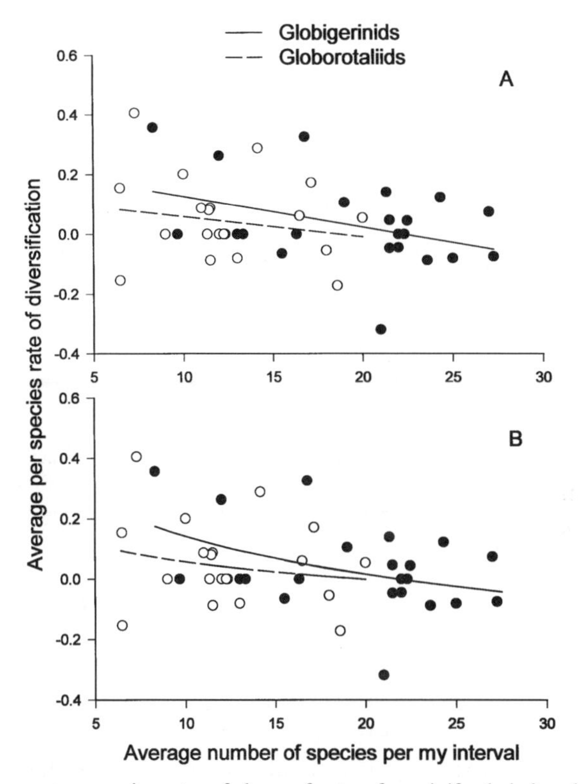

Fig. 9.3 Average per species rates of change for two foraminiferal clades plotted against average diversity of species. Curves represent the predicted relationship for the logistic (*A*) and Gompertz (*B*) models of species diversification (Maurer 1989). The per species rate of diversification when a single species is present was significantly greater for the ecologically generalized globigerinids than for the narrowly distributed globorotaliids. Globigerinids also had more species extant during the Neogene.

tion sizes for siphonate bivalves were much higher than for nonsiphonates primarily because siphonates were better able to escape predation. His conceptual model postulated that because siphonates had higher population densities, they were less likely to both speciate and become extinct than nonsiphonates. But the critical idea of his model was that the difference between these per species rates was greater for siphonates than nonsiphonates (fig. 9.4). In the two faunas Stanley studied, there were 116 siphonate and 24 nonsiphonate species. Although he did not comment on the difference in diversity of the two groups, the nearly five-fold difference between the groups suggests that the siphonates, with a greater difference between speciation and extinction rates, were also the most diverse. A couple of caveats, however, need to be mentioned. First, it is not clear that the siphonates and nonsiphonates represent sister clades.

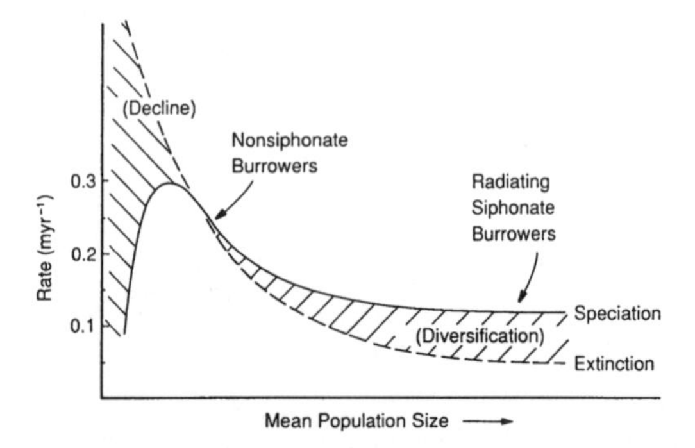

Fig. 9.4 Stanley's (1986) hypothetical model relating per species speciation and extinction rates for siphonate and nonsiphonate burrowing mollusks as a function of population size. Note that the siphonates have higher population sizes and lower speciation and extinction rates than the nonsiphonates, but the difference between per species speciation and extinction rates is greater for siphonates, the more diverse group. This model generates predictions regarding diversity and ecology that are similar to the statistical model developed in the text. (From Stanley 1990.)

The phylogeny for these groups is not presented with Stanley's data, so it is unclear whether we are dealing with groups of bivalves of similar ages. This may bias the comparison of diversities to favor siphonates if they generally tend to belong to older taxa. Use of sister groups precludes this kind of bias (Brooks and McLennan 1991). Second, since the assemblages considered may not include all species of each kind of bivalve, there may be geographic bias that implies that nonsiphonates are globally less diverse when they are not. Nevertheless, Stanley's data are suggestive.

Another interesting example which supports Darwin's view of diversification was given by Bleiweiss (1990). Bleiweiss examined the ecological characteristics of two sister clades of hummingbirds, the diverse (about 295 species) Trochilinae and the relatively depauperate Phaethornithinae (about 35 species). He found that species in Trochilinae were typically ecological generalists, with a generalized morphology and ability to disperse widely. On the other hand, the Phaethornithinae tended to be much more ecologically specialized and less vagile. Bleiweiss's argument for this apparent contradiction of the pattern suggested by macroevolutionists paralleled Darwin's general arguments described above. Since the Trochilinae are able to colonize and take advantage of novel ecological situations, they are more likely to speciate than the Phaethornithinae. Although no data were given on relative population sizes, it is likely that the Trochilinae are more common and have larger geographic ranges (excluding endemic island species). The paleontological record for birds is quite poor, and there is no way to determine per species rates of speciation and extinction for these birds. However, the data as they stand lend support to Darwin's scenario.

Conclusions

Species are generated during evolution with certain combinations of ecological attributes that allow them to achieve some degree of ecological success. Generally, species that are successful have relatively large geographic ranges and tend to have high population densities. Darwin foresaw that this would predispose such species to be more successful in competition with other species and to have a higher likelihood of persisting over geological time. In addition, any species that arose from an ecologically successful species would tend to have some of the same adaptations that would allow it also to be successful. Over time, Darwin argued that taxa comprising ecologically successful species would outlast and replace those that were less successful ecologically. Darwin was clearly thinking of what we now call clades, and his argument amounts to the claim that ecologically successful clades will be more diverse than related clades composed of species which have narrow ecological tolerances.

Some macroevolutionists have argued from a different perspective that ecologically successful species tend to be ecological generalists, and hence clades of ecological generalists tend to be species poor because being an ecological generalist tends to reduce the rates of speciation and extinction. But the analytical model developed above suggests that this perspective is partially mistaken. What is important is not the absolute rates of speciation and extinction, but the difference between these rates in the founding species of the clade. Since members of the clade inherit similar ecological attributes from their ancestors, the evolutionary conservatism of ecological traits implicit in the model tends to propagate ecological success or ecological inferiority. This translates eventually into characteristic differences in the overall rate of proliferation of species and, consequently, to differences in diversity. Stanley (1979, 1990) argued that speciation and extinction rates are generally correlated, and this may be so; but it is the magnitude of the difference between the rates that determines the diversity of a clade.

An important theoretical consequence of the view of taxonomic macroevolution developed in this chapter is that the phylogenetic history of a species has a major impact on its ecological success and its potential for subsequent diversification. Most species become extinct without giving rise to new species. The model developed in this chapter suggests that the species that do give rise to new clades are not a random sample of all existing species. Looking at the diversity of species living today, we would predict that, all else being equal, the species that are most successful are those that are likely to give rise to new forms, while more restricted species will become extinct without issue. This perspective also suggests that it may be fruitful to approach the study of modern ecology from a phylogenetic perspective (Brooks and McLennan 1991). Such an approach should complement the more reductionist studies common in ecology, particularly those that attempt to discover the processes responsible for generating species diversity (Ricklefs 1987).

Finally, note that a more explicit understanding of the constraints governing the dynamics of species diversity within a clade arose from adopting the statistical approach I have been suggesting. This is not to say that other approaches to the evolution of diversity will not provide valuable insights. However, I believe that large-scale problems, such as the evolution of diversity, are the kinds of problems where the statistical approach will be most successful. It is at geographic spatial scales and geological time scales that there are enough individuals contributing to the outcome of the processes of speciation and extinction to allow development of statistical models. Such models may gloss over some of the details that are important to problems like local community assembly, but the loss of detail is offset by the increased understanding of macroscopic processes such as speciation and extinction. In fact, macroscopic processes are relatively insensitive to what happens in specific instances at microscopic scales. Hence, much of the detail is probably irrelevant to macroscopic dynamics. Only microscopic processes that are constrained to occur in consistent spatial and temporal patterns (in a statistical sense) are relevant to macroscopic processes.

CHAPTER TEN

The Macroscopic Perspective and the Future of Ecology

The green prehuman earth is the mystery we were chosen to solve, a guide to the birthplace of our spirit.

E. O. Wilson (1992)

Despite the discoveries ecology has made in the past century, there is much we do not understand. The science of ecology is in its infancy. We have behind us some of the greatest thinkers modern science has produced, yet our vision of the future of our discipline is still in its formative stages. Despite the growing pains experienced by our science, we have a tremendous responsibility toward the global society to which we belong. The diverse paradigms that make up modern ecology are as a whole faced with a set of applied problems that, if not solved, may have profound consequences for our species and our world. There is thus a tremendous urgency to come up with a practical, workable science of ecology. Where, in this constantly evolving field, is the place for the macroscopic perspective that I have discussed throughout this book?

In this chapter I suggest three possible contributions that the macroscopic perspective alone can make in providing solutions to ecological problems. (There are undoubtedly other contributions and new perspectives in addition to these.) The first is mainly a philosophical contribution. The macroscopic perspective of community ecology reinforces the idea that all ecological systems are ultimately connected, no matter how remote they are from one another. This has profound implications. Any ecological system humans attempt to manipulate or study will send out ripples of causation that affect other, distantly located ecosystems. The reverse is also true: it will seldom be possible to understand any local system without knowing its context. The second contribution is a practical one. Learning what we can from the macroscopic perspective may help us set conservation priorities and goals when we are in a situation where nearly every system is in trouble. Finally, the macroscopic effort suggests that ecologists' statistical toolbox is badly in need of augmentation. There are statistical techniques out there that are capable of handling more complicated data and answering more complex questions. Unfortunately, such techniques are not commonly taught in a one-year course in statistics.

No Ecosystem Is an Island

Over ten years ago, Jim Brown and I went to tour the Biosphere 2 site when it was in its planning stages. We were both impressed with the enormity of the undertaking that was being planned. A great deal of research and work went into selecting which particular species to place in the enclosure. Near the end of our tour our gracious hosts took us to a model of the planned experiment. My first impression was "Isn't this structure going to be too small to support eight people?" I do not remember if I brought the point up during the discussion with our hosts, but at that time I only had a vague notion of why I was uncomfortable with the size of the experimental structure.

A few years later, the structure was complete, and, with a bit of fanfare, the eight participants were locked up with small rain forest, coral reef, marsh, savanna, and desert habitats, along with a relatively small plot for intensive agriculture. Everyone involved in the experiment was convinced that Biosphere 2 would work. Yet after the experiment had run for a short while, a shorter time than initially expected, it was clear that Biosphere 2 was not working like it was supposed to work. Carbon dioxide levels increased to a point that infusions of air were required. Of the 300 species of plants initially introduced into the rain forest habitat, 100 survived. Extinctions occurred in other habitats as well. Although some might view the behavior of Biosphere 2 as a disaster, I view it as a success. It taught us something important about our own biosphere.

One reason that Biosphere 2 did not behave as expected was that it did not replicate the atmospheric component of our own biosphere very well. For every square meter of the earth's surface, there are about 50 cubic meters of atmosphere. Biosphere 2, however, had only 13 cubic meters of atmosphere for every square meter of its surface. Such a small atmospheric ratio led to an atmosphere that was not able to sustain an ecosystem that could support eight humans (and viable populations of many of the species of plants and animals, as well).

I believe there was something else wrong with Biosphere 2. I do not think it was big enough. The reason has to do with the ability of a complex system to withstand fluctuations. Consider the following analogy. Imagine that two different fast-food companies are attempting to compete in a newly opened market. Suppose one of them was formed by a local entrepreneur, while the second was opened by a large, multinational corporation. All else being equal, the store started by the large corporation will have a greater probability of being successful because it will have an easier time weathering slowdowns in the local economy. This is because each store for the large corporation need make a profit only for a relatively short period of time. When the profits and losses of many individual stores are added together in the books of the parent corporation, the earnings of any one store will have little effect on the entire corporation. The small, locally owned store must make a profit in order to survive. It has no resources upon which to draw during lean times in the local economy.

An ecosystem is a lot like a store embedded in the economy of nature. An isolated ecosystem must make all energy flow and nutrient cycle "budget sheets" balance. If it does not, the ecosystem will lose parts (i.e., populations of some species will become extinct). But local ecosystems connected to the biosphere can "ride out" times when there might be a net energy loss or nutrient deficiency because these will eventually be augmented by flows from connected ecosystems. Many populations may operate in a similar manner. Recall the model of geographic range structure considered in chapter 7. By being part of a larger geographic range, a local population may only need to experience positive net growth once in a while to be a positive contributor to the total population of the entire species.

This view of populations, communities, and ecosystems of persisting by virtue of their connections to larger systems has important implications for the way that we interact as a species with the biosphere. Many of our activities can be construed as decreasing the size of the biosphere available to other species. In doing so, we make individual species or ecosystems less likely to persist, and consequently decrease the diversity within the biosphere. In the next section I consider how the macroscopic perspective might contribute to biological conservation.

Persistence of Diversity in a Shrinking Biosphere

Biosphere 2 shrank to fit its limitations. It is considerably simpler and less diverse now than it was when it was first constructed. Although it is no longer a closed system, it still cannot maintain the kind of diversity that the original designers hoped it would. What seems very clear now is that human activities are using up a large amount of the resources that once maintained the biosphere (Vitousek et al. 1986, 1997; Wright 1990; Vitousek 1994). With respect to most other species, the biosphere is shrinking. For that matter, it is shrinking for our own species, but we have always been able to use technology to offset losses due to that

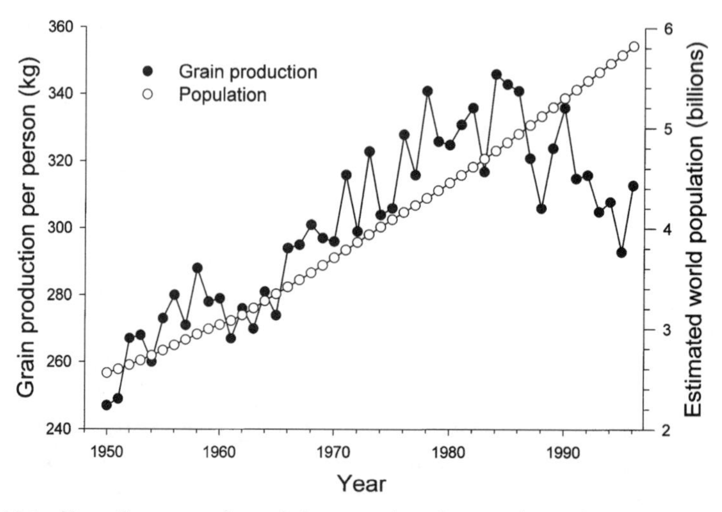

Fig. 10.1 Forty-five years of population growth and per capita grain production. Note the downturn in per capita grain production that began in 1984. Although there have been some yearly increases, the overall trend is a decline.

shrinkage. There are signs, however, that suggest that this may not always be possible. For example, the per capita production of grain has been declining since 1984 (fig. 10.1). Like Biosphere 2, our own biosphere is shrinking to fit the limitations that we are imposing upon it.

One of the most critical problems we face in maintaining a biosphere that can support a viable human population is maintaining sufficient biological diversity to ensure that the services ecosystems provide for humans are not compromised (Wilson 1992; Ehrlich 1997). A relatively simple calculation illustrates this problem. Given rough estimates of human per capita consumption (this includes energy diverted to agriculture), I estimated that a human population of about 20–30 billion would use essentially all primary productivity to support itself and its agricultural interests (Maurer 1996). Assuming that human consumption does not change, the time frame in which this will occur is about two hundred years at current rates of human population growth (fig. 10.2). This estimate would be considerably lengthened if human population growth were slowed to about 10% of its current rate of increase of 1.7% per year. The uncertainty in these kinds of calculations is substantial (Cohen 1995), but even in a best-case scenario there appears to be a certain decrease in the biological diversity that will result from increasing human demands on the biosphere.

There is a clear sense of urgency in conservation biology that feeds

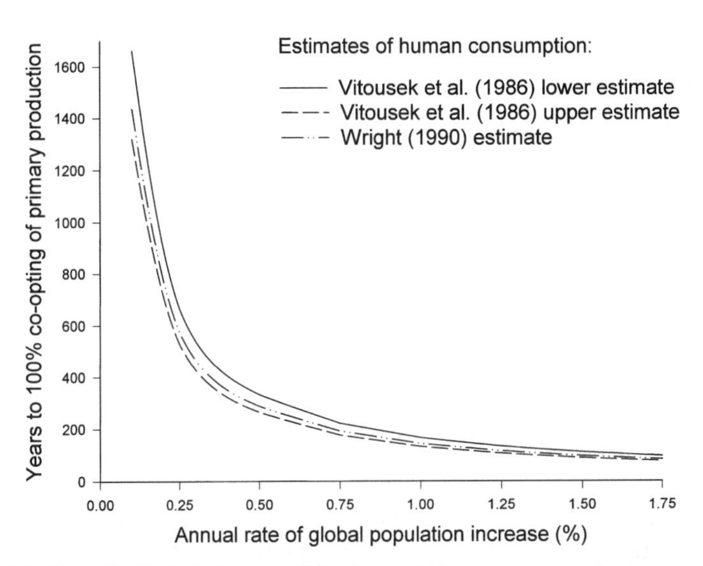

Fig. 10.2 Hypothetical calculations of the time until humans co-opt all primary productivity for themselves and their agricultural activities as a function of the rate of growth of the human population. Modified from Maurer 1996.

upon the kinds of calculations we can make about the future of the world ecosytem. But there is also a strong sense of stewardship. We are presenting a very different kind of biosphere to future generations than was given to us. The generations of humans now living were entrusted with a world that was full of essential resources for human use. Although somewhat depleted by the industrial revolution, the resources that were available when I was born were still expansive. Our generation, however, by allowing human consumption and population to increase without being checked, is beginning to erode the very foundation of our prosperity.

The macroscopic perspective cannot by itself provide the sweeping solutions to conservation problems that are needed. But it can help. In discussing political aspects of environmental management, Gore (1992) differentiated between tactical approaches and strategic approaches. Tactical approaches provide specific solutions to specific problems. The goal of this kind of conservation is to mobilize political and economic resources to provide ecologically sound management tactics in reaction to specific changes induced in local ecosystems. Strategic approaches, on the other hand, provide criteria for prioritizing tactical actions and rationale for implementing one set of tactics instead of another given the class of problem to be solved. Tactical conservation has a firm theoretical foundation in existing ecological concepts (e.g., Caughley and Gunn 1996). There is a less well developed foundation for strategic conservation practices. Most often, strategic decisions are influenced by immediate economic goals, the idiosyncrasies of individual conservation interests (Clark 1997), or other similarly nonscientific considerations. This is where macroecology might be able to help (Brown 1995). For example, suppose a government agency is given limited resources to conserve the ecosystems that will be most likely to maintain biological diversity for a long period of time. Which areas should be preserved? Which species?

One approach to answering this question has been the gap analysis program (GAP; Scott et al. 1993). The general idea was to provide guidance to political entities (states) in selecting the best geographic regions for conservation priority. The approach was to identify regions within a state that had the highest species diversity, and to compare these with land ownership patterns. The state could then prioritize which geographic regions to focus its resources on in order to maximize its ability to preserve biological diversity. For example, Utah's gap analysis program identified three regions in the state that had unique species in them and that were not currently covered by federal or state conservation programs (T. Edwards, personal communication). Of the three, only the southwest corner of the state had sufficient federal or state lands to provide a realistic opportunity to improve conservation efforts. Despite these data, state officials chose to prioritize economic development ahead of conservation interests in southwestern Utah.

The power of gap analysis is that it provides a strategic planning tool for conservation. Despite some concerns about data quality (Short and Hestbeck 1995), tests of predictions made from gap analysis programs are encouraging (Edwards et al. 1996). Data quality is not a major concern, in my mind, about the gap analysis program. My concern stems from the nature of regions that have high species richness within a landscape. High richness may come from two different sources. First, it might be due to relatively high productivity of a particular landscape, within which viable populations of many species might be found. Second, it might occur within a landscape because of the juxtaposition of a wide variety of different habitat types. Species may not be able to maintain viable populations within each habitat type in the landscape, but rather persist there due to frequent immigration from elsewhere. That is, a high-diversity landscape may be composed of a high diversity of habitat sinks. Gap analysis, as currently practiced, does not differentiate between these two alternatives. An example of this is the rare plant species in Glacier National Park (T. Williams, personal communication). Of over 50 species identified as rare within the park, only one is endemic. All the rest are species for which Glacier National Park is at the edge of their geographic range. Although there is much that is desirable in maintaining

Glacier National Park as a conservation reserve, it is unlikely that it would provide sufficient protection for biological diversity on its own.

From considerations of geographic range structure, a somewhat different perspective is added to species conservation. One of the insights from considering geographic range structure is that the species most likely to be responsible for generating the most diversity in the future tend to be widespread and abundant species. Much conservation is focused on saving species that are rare, or endangered. In some cases, rarity is a recent condition imposed upon a species by drastic changes in the environment induced by human activities. In other cases, rarity is probably a condition that has always been experienced by the species. Although there are many reasons to focus conservation efforts on rare species, one guideline from the macroscopic perspective is that a once-widespread species that has declined to rarity is more indicative of major disruptions of the environment than the rarity of a species that was rare before humans occupied the landscape.

Expanding the Ecological Tool Kit

In attempting to answer complex questions, as is required of community ecologists of both the applied and the theoretical persuasions, it is no longer sufficient to rely on the simple statistical tests that were used by previous generations of ecologists. In some instances, a simple *t*-test might still be the best tool for the job, but reliance on such statistics cannot answer the more complicated questions that we must be able to answer to progress in ecology. Perhaps more dangerous than the wrong answers that might be given in some situations is the false sense of rigor that imbues ecological investigations by acceptance or rejection of dichotomous hypotheses (Hilborn and Mangel 1997). Although the choice between hypotheses is unambiguous in conventional statistical practice, hypotheses themselves may be ill posed. In a complex system, there may be many synergistic parameters that are simultaneously different between systems being studied. This means that there may be several competing models, each of which can explain the data to a certain degree. Choosing among several such models is no job for a *t*-test.

Fortunately, there are alternatives. Hilborn and Mangel (1997) discussed the maximum likelihood principle, and how it can be used to choose between different, competing models. The problem with maximum likelihood is that there are no preset, simple formulas to use. It is a sophisticated statistical tool that must be appropriately tailored to specific problems. The user must be able not only to apply the statistical technique properly but also to formulate appropriate ecological models to test. This means that ecologists will need to be trained in more sophisticated quantitative skills.

Consider the following analysis to see the power of the general technique of testing alternative models. Lele, Taper, and Gage (1998) modeled the dispersal of gypsy moths (*Lymantria dispar*) in Michigan from 1985 to 1994. The gypsy moth is an aggressive colonist that has done a great deal of damage to native forests since its introduction into Massachusetts in 1869. Understanding and being able to predict its population dynamics in space is an important part of conservation. Lele, Taper, and Gage examined four different models of the dynamics of gypsy moth populations in Michigan. Their first alternative was a model with density-dependent growth at individual sites. Dispersal was not included in the model, but correlations in density among different sites were assumed to be due to spatial autocorrelation in environmental conditions. Density-independent growth was assumed to be the same across all sites. This was essentially a "null model" relative to two different effects: (1) spatial variation in the density-independent component of population growth, and (2) no population effects of dispersal among sites. The second model was the same as the first but added individual density-independent growth terms for each site. It tested the effect of spatial variation on density independence. The third model was identical to the first, except it added a term for dispersal among sites into the population growth model. This model tested the effect of dispersal. Finally, the fourth model included terms for both spatial variation in density independence and dispersal. I will not write out the equations for the model here; the interested reader is referred to Lele, Taper, and Gage 1998 for the details. The technique that the authors used to estimate parameters is related to maximum likelihood estimation procedures, and gave them unbiased estimates of the parameters of their complex models.

The interesting result that Lele, Taper and Gage (1998) found was that models that included dispersal did better than those that did not (fig. 10.3). The inclusion of individual terms for density independence at each site, however, did not seem to improve predictions a great deal. Thus, for the gypsy moth in Michigan, one can tentatively conclude that populations at each site do not differ a great deal in the way they react to density-independent factors. It is likely, though, that they are tied together by dispersal from one site to the next. Thus, population dynamics of gypsy moths across space are autocorrelated for two reasons: (1) populations are tied together by dispersal, and (2) populations are living in an autocorrelated environment.

The advantages of approaches like the one described above are many. They provide a mechanistic understanding of the processes that deter-

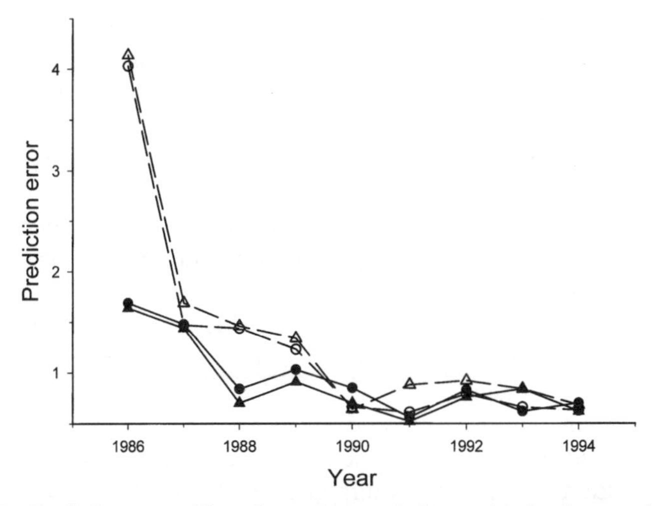

Fig. 10.3 Prediction error of four alternative population models for the spread of gypsy moths in Michigan. *Closed symbols* represent models that include dispersal, *open symbols* models that do not. *Circles* represent models with uniform density-independent terms, *triangles* represent models that have spatially varying density independence. A lower value indicates a better model (lower prediction error).

mine larger-scale spatial patterns. There are explicit connections between lower-level processes, such as dispersal and density dependence, and higher-level spatial patterns. They are also tractable statistically, so that parameter estimates can be obtained. A major advantage over simple hypothesis testing is that such modeling techniques provide objective criteria for evaluating multiparameter hypotheses like spatial variation in density dependence and dispersal. Some ecologists might protest, asking questions like "Where are the *significance tests*?" Reliance on significance tests, however, gives one a false sense of assurance in one's results. Significance levels are only estimates, after all, based on sometimes faulty assumptions. The criterion for model evaluation in the example above was the ability to predict the data accurately. The models do not explain everything about the data, but they certainly allow one to evaluate objectively alternative hypotheses about processes.

Conclusions

The macroscopic perspective that I have described in the latter chapters of this book has much to offer ecology. Given the understanding of ecological phenomena like competition and predation that has been obtained through intensive, local-scale, often experimental studies, the macroscopic approach to ecology provides a complementary view of ecology that gives us the opportunity to add significantly to our knowledge about communities and ecosystems. In the future, the dialogue between "holists" and "reductionists" in ecology should be positive and interactive. Experiments can provide key insights into specific processes that might be important in determining the geographic distribution of species. They can provide information about what forms mechanistic equations for spatio-temporal population dynamics might take. Large-scale geographic studies will provide information about the context within which the dynamics of local communities and ecosystems occur. They can provide additional insights into specific hypotheses that might be tested experimentally.

A cross-pollination of macro and micro studies in ecology has the potential to provide useful answers to complex ecological questions. Understanding patterns and processes at multiple scales, after all, is what ecologists have been calling for recently (Lubchenco et al. 1991; Levin 1992; Root and Schneider 1995). Finding answers to complex ecological questions is extremely important if ecology is to help society react in a constructive manner to the effects of a shrinking biosphere.

LITERATURE CITED

Abrams, P. 1993. Why predation rate should not be proportional to predator density. *Ecology* 74:726–33.

Allen, T. F. H., and T. W. Hoekstra. 1992. *Toward a unified ecology.* New York: Columbia University Press.

Allen, T. F. H., R. V. O'Neill, and T. H. Hoekstra. 1987. Interlevel relations in ecological research and management: Some working principles from hierarchy theory. *Journal of Applied Systems Analysis* 14:63–79.

Allen, T. F. H., and T. B. Starr. 1982. *Hierarchy: Perspectives for ecological complexity.* Chicago: University of Chicago Press.

American Ornithologists' Union (AOU). 1957. *Check-list of North American birds.* 5th ed. Ithaca, NY: American Ornithologists' Union.

———. 1983. *Check-list of North American birds.* 6th ed. Ithaca, NY: American Ornithologists' Union.

Anderson, G. R. V., A. H. Ehrlich, P. R. Ehrlich, J. D. Roughgarden, B. C. Russell, and F. H. Talbot. 1981. The community structure of coral reef fishes. *American Naturalist* 117:476–95.

Andrewartha, H. G., and L. C. Birch. 1984. *The ecological web: More on the distribution and abundance of animals.* Chicago: University of Chicago Press.

Ansell, W. F. H. 1971. Part 15: Order Artiodactyla. In *The mammals of Africa: An identification manual,* ed. J. Meester and H. W. Setzer, 11–84. Washington, D.C.: Smithsonian Institution Press.

Atmar, W., and B. D. Patterson. 1993. The measure of order and disorder in the distribution of species in fragmented habitat. *Oecologia* 96:373–82.

Barahona, M., and C. Poon. 1996. Detection of non-linear dynamics in short, noisy time series. *Nature* 381:215–17.

Baskin, Y. 1997. Center seeks synthesis to make ecology more useful. *Science* 275:310–11.

Begon, M., J. L. Harper, and C. R. Townsend. 1990. *Ecology: Individuals, populations and communities.* 2d ed. Boston: Blackwell Scientific Publications.

Bender, E. A., T. J. Case, and M. E. Gilpin. 1984. Perturbation experiments in community ecology: Theory and practice. *Ecology* 65:1–13.

Billick, I., and T. J. Case. 1994. Higher order interactions in ecological communities: What are they and how can they be detected? *Ecology* 75:1529–43.

Blackburn, T. M., and K. J. Gaston. 1994. The distribution of body sizes of the world's bird species. *Oikos* 70:127–30.

Blackburn, T. M., P. H. Harvey, and M. D. Pagel. 1990. Species number, population density, and body size relationships in natural communities. *Journal of Animal Ecology* 59:335–45.

Blackburn, T. M., J. H. Lawton, and S. L. Pimm. 1993. Non-metabolic explanations for the relationship between body size and animal abundance. *Journal of Animal Ecology* 62:694–702.
Bleiweiss, R. 1990. Ecological causes of clade diversity in hummingbirds: A neontological perspective on the generation of diversity. In *Causes of evolution: A paleontological perspective,* ed. R. M. Ross and W. D. Allmon, 354–80. Chicago: University of Chicago Press.
Bock, C. E. 1984. Geographical correlates of abundance vs. rarity in some North American winter landbirds. *Auk* 101:266–73.
———. 1987. Distribution-abundance relationships of some Arizona land birds: A matter of scale? *Ecology* 68:124–29.
Bock, C. E., and R. E. Ricklefs. 1983. Range size and local abundance of some North American songbirds: A positive correlation. *American Naturalist* 122: 295–99.
Bonner, J. T. 1988. *The evolution of complexity by means of natural selection.* Chicago: University of Chicago Press.
Bowers, J. H., R. J. Baker, and M. H. Smith. 1973. Chromosomal, electrophoretic, and breeding studies of selected populations of deer mice (*Peromyscus maniculatus*) and black-eared mice *(P. melanotis). Evolution* 27:378–86.
Bradley, R. A. 1983. Complex food webs and manipulative experiments in ecology. *Oikos* 41:150–52.
Bray, J. R., and J. T. Curtis. 1957. An ordination of the upland forest communities of southern Wisconsin. *Ecological Monographs* 27:325–49.
Brian, M. V. 1953. Species frequencies in random samples from animal populations. *Journal of Animal Ecology* 22:57–64.
Brooks, D. R. 1985. Historical ecology: A new approach to studying the evolution of ecological associations. *Annals of the Missouri Botanical Gardens* 72:660–80.
Brooks, D. R., and D. A. McLennan. 1991. *Phylogeny, ecology, and behavior.* Chicago: University of Chicago Press.
Brooks, D. R., and E. O. Wiley. 1988. *Evolution as entropy: Toward a unified theory of biology.* Chicago: University of Chicago Press.
Brown, J. H. 1981. Two decades of homage to Santa Rosalia: Toward a general theory of diversity. *American Zoologist* 21:877–88.
———. 1984. On the relationship between distribution and abundance. *American Naturalist* 124: 255–79.
———. 1995. *Macroecology.* Chicago: University of Chicago Press.
———. 1997. An ecological perspective on the challenge of complexity. EcoEssay Series no. 1. National Center for Ecological Analysis and Synthesis, Santa Barbara. CA.
Brown, J. H., D. W. Davidson, J. C. Munger, and R. S. Inyoue. 1986. Experimental community ecology: The desert granivore system. In *Community ecology,* ed. J. Diamond and T. J. Case, 41–61. New York: Harper & Row.
Brown, J. H., and A. C. Gibson. 1983. *Biogeography.* St. Louis: Mosby.
Brown, J. H., and E. J. Heske. 1990. Control of a desert-grassland transition by a keystone rodent guild. *Science* 250:1705–7.
Brown, J. H., and M. A. Kurzius. 1987. Composition of desert rodent faunas: Combinations of coexisting species. *Annales Zoologici Fennici* 24:227–37.

———. 1989. Spatial and temporal variation in guilds of North American granivorous rodents. In *Patterns in the structure of mammalian communities,* ed. D. W. Morris, Z. Abramsky, B. J. Fox, and M. R. Willig, 71–90. Special Publications of the Museum, no. 28. Lubbock: Texas Tech. University Press.

Brown, J. H., P. A. Marquet, and M. L. Taper. 1993. Evolution of body size: Consequences of an energetic definition of fitness. *American Naturalist* 142: 573–84.

Brown, J. H., and B. A. Maurer. 1986. Body size, ecological dominance and Cope's rule. *Nature* 324:248–51.

———. 1987. Evolution of species assemblages: Effects of energetic constraints and species dynamics on the diversification of the North American terrestrial avifauna. *American Naturalist* 130:1–17.

———. 1989. Macroecology: The division of food and space among species on continents. *Science* 243:1145–50.

Brown, J. H., D. W. Mehlman, and G. C. Stevens. 1995. Spatial variation in abundance. *Ecology* 76:2028–43.

Brown, J. H., and J. C. Munger. 1985. Experimental manipulation of a desert rodent community: Food addition and species removal. *Ecology* 66:1545–63.

Brown, J. H., and P. F. Nicoletto. 1991. Spatial scaling of species composition: Body masses of North American land mammals. *American Naturalist* 138: 1478–512.

Brown, J. H., M. L. Taper, and P. A. Marquet. 1996. Darwinian fitness and reproductive power: Reply to Kozlowski. *American Naturalist* 147:1092–97.

Brown, J. H., T. J. Valone, and C. G. Curtin. 1997. Reorganization of an arid ecosystem in response to recent climate change. *Proceedings of the National Academy of Sciences, USA* 94:9729–33.

Brown, J. H., and Z. Zeng. 1989. Comparative population ecology of eleven species of rodents in the Chihuahuan Desert. *Ecology* 70:1507–25.

Browne, J. 1980. Darwin's botanical arithmetic and the "principle of divergence," 1854–1858. *Journal of the History of Biology* 13:53–89.

Calder, W. A., III. 1984. *Size, function, and life history.* Cambridge, MA: Harvard University Press.

Carpenter, S. R., S. W. Chisholm, C. J. Krebs, D. W. Schindler, and R. F. Wright. 1995. Ecosystem experiments. *Science* 269:324–27.

Caswell, H., and J. E. Cohen. 1993. Local and regional regulation of species-area relations: A patch occupancy model. In *Species diversity in ecological communities,* ed. R. E. Ricklefs and D. Schluter, 99–107. Chicago: University of Chicago Press.

Caughley, G., and A. Gunn. 1996. *Conservation biology in theory and practice.* Cambridge, MA: Blackwell Science.

[Chambers, Robert]. 1844. *Vestiges of the natural history of creation.* London: John Churchill.

Charnov, E. L. 1993. *Life history invariants.* Oxford: Oxford University Press.

Chatfield, C. 1996. *The analysis of time series: An introduction.* London: Chapman & Hall.

Clark, T. W. 1997. *Averting extinction.* New Haven, CT: Yale University Press.

Clements, F. E. 1916. *Plant succession.* Publication no. 242. Washington, D.C.: Carnegie Institute of Washington.

———. 1936. Nature and structure of the climax. *Journal of Ecology* 24:252–84.

Cohen, J. E. 1995. *How many people can the earth support?* New York: W. W. Norton.

Coleman, B. D., M. A. Mares, M. R. Willig, and Y.-H. Hsieh. 1982. Randomness, area, and species richness. *Ecology* 63:1121–33.

Collier, J. 1988. The dynamics of biological order. In *Entropy, information, and evolution: New perspectives on physical and biological evolution,* ed. B. H. Weber, D. J. Depew, and J. D. Smith, 227–42. Cambridge, MA: MIT Press.

Collins, S. L., and S. M. Glenn. 1990. A hierarchical analysis of species' abundance patterns in grassland vegetation. *American Naturalist* 135:633–48.

Colwell, R. K., and G. C. Hurtt. 1994. Nonbiological gradients in species richness and a spurious Rapport effect. *American Naturalist* 144:570–95.

Connell, J. H. 1983. On the prevalence and relative importance of interspecific competition: Evidence from field experiments. *American Naturalist* 122:661–96.

Connor, E. F., and E. D. McCoy. 1979. The statistics and biology of the species-area relationship. *American Naturalist* 113:791–833.

Connor, E. F., and D. Simberloff. 1978. Species number and compositional similarity of the Galapagos flora and avifauna. *Ecological Monographs* 48:219–48.

———. 1979. The assembly of communities: Chance or competition? *Ecology* 60:1132–40.

Conrad, M. 1983. *Adaptability: The significance of variability from molecule to ecosystem.* New York: Plenum Press.

Cornell, H. V., and J. H. Lawton. 1992. Species interactions, local and regional processes and limits to the richness of ecological communities: A theoretical perspective. *Journal of Animal Ecology* 61:1–12.

Cowles, H. C. 1899. The ecological relations of the vegetation on the sand dunes of Lake Michigan. *Botanical Gazette* 27:95–117, 167–202, 281–308, 361–91.

Crowell, K. L., and S. L. Pimm. 1976. Competition and niche shifts of mice introduced onto small islands. *Oikos* 27:442–54.

Curnutt, J. C., S. L. Pimm, and B. A. Maurer. 1996. Population variability of sparrows in space and time. *Oikos* 76:131–44.

Damuth, J. 1981. Population density and body size in animals. *Nature* 290:699–700.

———. 1987. Interspecific allometry of population density in mammals and other animals: The independence of body mass and population energy-use. *Biological Journal of the Linnean Society* 31:193–246.

Darwin, C. 1859. *On the origin of species by means of natural selection.* London: John Murray.

Davidson, D. W., D. A. Samson, and R. S. Inouye. 1985. Granivory in the Chihuahuan Desert: Interactions within and between trophic levels. *Ecology* 66: 486–502.

Delcourt, H. R., and P. A. Delcourt. 1991. *Quaternary ecology.* London: Chapman & Hall.

Dial, K. P., and J. M. Marzluff. 1988. Are the smallest organisms the most diverse? *Ecology* 69:1620–24.

Diamond, J. M. 1975. Assembly of species communities. In *Ecology and evolution of communities,* ed. M. L. Cody and J. M. Diamond, 342–44. Cambridge, MA: Harvard University Press.

———. 1978. Niche shifts and rediscovery of interspecific competition. *American Scientist* 66:322–31.
———. 1986. Overview: Laboratory experiments, field experiments, and natural experiments. In *Community ecology*. ed. J. Diamond and T. J. Case, 3–22. New York: Harper & Row.
Dobson, A. J. 1983. An introduction to statistical modelling. London: Chapman & Hall.
Drake, J. A. 1990. Communities as assembled structures: Do rules govern pattern? *Trends in Ecology and Evolution* 5:159–64.
Dunning, J. B. 1993. *CRC handbook of avian body masses.* Boca Raton, FL: CRC Press.
Dunson, W. A., and J. Travis. 1991. The role of abiotic factors in community organization. *American Naturalist* 138:1067–91.
Edelstein-Keshet, L. 1988. *Mathematical models in biology.* New York: Random House.
Edwards, T. C., Jr., E. T. Deshler, D. Foster, and G. G. Moisen. 1996. Adequacy of wildlife habitat relation models for estimating spatial distributions of terrestrial vertebrates. *Conservation Biology* 10:263–70.
Efron, B., and R. J. Tibshirani. 1993. *An introduction to the bootstrap.* New York: Chapman & Hall.
Ehrlich, P. 1997. *A world of wounds: Ecologists and the human dilemma.* Olendorf/Luhe, Germany: Ecological Institute.
Eldredge, N. 1979. Alternative approaches to evolutionary theory. *Bulletin of the Carnegie Museum of Natural History* 13:7–19.
———. 1985. *Unfinished synthesis: Biological hierarchies and modern evolutionary thought.* New York: Oxford University Press.
———. 1989. *Macroevolutionary dynamics.* New York: McGraw-Hill.
Eldredge, N., and S. J. Gould. 1972. Punctuated equilibria: An alternative to phyletic gradualism. In *Models in paleobiology,* ed. T. J. M. Schopf, 82–115. San Francisco, CA: Freeman, Cooper & Co.
Eldredge, N., and S. N. Salthe. 1984. Hierarchy and evolution. In *Oxford surveys in evolutionary biology,* ed. R. Dawkins and M. Ridley, 182–206. Oxford: Oxford University Press.
Ellner, S., and P. Turchin. 1995. Chaos in a noisy world: New methods and evidence from time-series analysis. *American Naturalist* 145:343–75.
Emlen, J. M. 1984. *Population biology: The coevolution of population dynamics and behavior.* New York: Macmillan.
Emlen, J. T. 1978. Density anomalies and regulatory mechanisms in land bird populations on the Florida peninsula. *American Naturalist* 112:265–68.
Emlen, J. T., M. J. DeJong, M. J. Jaeger, T. C. Moermond, K. A. Rusterholz, and R. P. White. 1986. Density trends and range boundary constraints of forest birds along a latitudinal gradient. *Auk* 103:791–803.
Endler, J. A. 1986. Natural selection in the wild. Princeton, NJ: Princeton University Press.
Ewens, W. J., P. J. Brockwell, J. M. Gani, and S. I. Resnick. 1987. Minimum viable population size in the presence of catastrophes. In *Viable populations for conservation,* ed. M. E. Soule, 59–68. Cambridge: Cambridge University Press.

Farlow, J. O. 1976. A consideration of the trophic dynamics of a late Cretaceous large-dinosaur community (Oldman Formation). *Ecology* 57:841–57.
Fisher, R. A. 1958. *The genetical theory of natural selection.* Oxford: Clarendon Press.
Fisher, R. A., A. S. Corbet, and C. B. Williams. 1943. The relation between the number of species and the number of individuals in a random sample of an animal population. *Journal of Animal Ecology* 12:42–58.
Forbes, S. A. 1887. The lake as a microcosm. *Bulletin of the Peoria Scientific Association,* 77–87.
Forman, R. T. T., and M. Godron. 1986. *Landscape ecology.* New York: Wiley.
Frautschi, S. 1988. Entropy in an expanding universe. In *Entropy, information, and evolution: New perspectives on physical and biological evolution,* ed. B. H. Weber, D. J. Depew, and J. D. Smith, 11–22. Cambridge, MA: MIT Press.
Garland, T., Jr., A. W. Dickerman, C. M. Janis, and J. A. Jones. 1993. Phylogenetic analysis of variance by computer simulation. *Systematic Biology* 42:265–92.
Gaston, K. J. 1988. Patterns in the local and regional dynamics of moth populations. *Oikos* 53:49–57.
———. 1994. *Rarity.* New York: Chapman & Hall.
Gaston, K. J., and T. M. Blackburn. 1996a. Global scale macroecology: Interactions between population size, geographic range size, and body size in the Anseriformes. *Journal of Animal Ecology* 65:701–14.
———. 1996b. Range size–body size relationships: Evidence of scale dependence. *Oikos* 75:479–85.
Gaston, K. J., T. M. Blackburn, and J. H. Lawton. 1993. Comparing animals and automobiles: A vehicle for understanding body size and abundance relationships in species assemblages? *Oikos* 66:172–79.
Gaston, K. J., and J. H. Lawton. 1989. Insect herbivores on bracken do not support the core-satellite hypothesis. *American Naturalist* 134:761–77.
Gauch, H. G. 1982. *Multivariate analysis in community ecology.* Cambridge: Cambridge University Press.
Gause, G. F. 1930. Studies on the ecology of the Orthoptera. *Ecology* 11:307–25.
———. 1931. The influence of ecological factors on the size of population. *American Naturalist* 65:70–76.
———. 1932. Ecology of populations. *Quarterly Review of Biology* 7:27–46.
———. 1934. *The struggle for existence.* Baltimore, MD: Williams & Wilkins.
Gilpin, M. E. 1988. A comment on Quinn and Hastings: Extinction in subdivided habitats. *Conservation Biology* 2:290–92.
Gilpin, M. E., M. P. Carpenter, and M. J. Pomerantz. 1986. The assembly of a laboratory community: Multispecies competition in Drosophila. In *Community ecology,* ed. J. Diamond and T. J. Case, 23–40. New York: Harper & Row.
Gittleman, J. L., and M. Kot. 1990. Adaptation: Statistics and a null model for estimating phylogenetic effects. *Systematic Zoology* 39: 227–41.
Gittleman, J. L., and H.-K. Luh. 1992. On comparing comparative methods. *Annual Review of Ecology and Systematics* 23:383–404.
Gleason, H. A. 1926. The individualistic concept of the plant association. *Bulletin of the Torrey Botanical Club* 53:7–26.

Gleick, J. 1987. *Chaos: Making a new science.* New York: Penguin Books.

Gonzalez, M. J., and T. M. Frost. 1994. Comparisons of laboratory bioassays and a whole-lake experiment: Rotifer responses to experimental acidification. *Ecological Applications* 4:69–80.

Goodman, D. 1987. The demography of chance extinction. In *Viable populations for conservation,* ed. M. E. Soule, 11–34. Cambridge: Cambridge University Press.

Gore, A. 1992. *Earth in the balance.* New York: Penguin Books.

Gotelli, N. J., and G. R. Graves. 1996. *Null models in ecology.* Washington, D.C.: Smithsonian Institution Press.

Gotelli, N. J., and D. Simberloff. 1987. The distribution and abundance of tallgrass prairie plants: A test of the core-satellite hypothesis. *American Naturalist* 130:18–35.

Gould, S. J. 1988. Trends as changes in variance: A new slant on progress and directionality in evolution. *Journal of Paleontology* 62:319–29.

———. 1989. *Wonderful life.* New York: Norton.

Gould, S. J., and N. Eldredge. 1977. Punctuated equilibrium: The tempo and mode of evolution reconsidered. *Paleobiology* 3:115–51.

Graham, R. W. 1986. Response of mammalian communities to environmental changes during the late Quaternary. In *Community ecology,* ed. J. Diamond and T. J. Case, 300–313. New York: Harper & Row.

Greenacre, M. J., and E. S. Vrba. 1984. Graphical display and interpretation of antelope census data in African wildlife areas, using correspondence analysis. *Ecology* 65:984–97.

Grinnell, J. 1917. The niche-relationships of the California thrasher. *Auk* 34: 427–33.

Gurevitch, J., and L. V. Hedges. 1993. Meta-analysis: Combining the results of independent experiments. In *Design and analysis of ecological experiments,* ed. S. M. Scheiner and J. Gurevitch, 378–98. New York: Chapman & Hall.

Gurevitch, J., L. L. Morrow, A. Wallace, and J. Walsh. 1992. A meta-analysis of competition in field experiments. *American Naturalist* 140:539–72.

Gyllenberg, M., I. Hanski, and A. Hastings. 1997. Structured metapopulation models. In *Metapopulation biology,* ed. I. Hanski and M. E. Gilpin, 93–122. San Diego, CA: Academic Press.

Hairston, N. G., Sr. 1989. *Ecological experiments: Purpose, design, and execution.* Cambridge: Cambridge University Press.

Hairston, N. G., F. E. Smith, and L. B. Slobodkin. 1960. Community structure, population control, and competition. *American Naturalist* 91:421–25.

Hallet, J. G., and S. L. Pimm. 1979. Direct estimation of competition. *American Naturalist* 113:593–600.

Hanski, I. 1982a. Dynamics of regional distribution: The core and satellite species hypothesis. *Oikos* 38:210–21.

———. 1982b. Communities of bumblebees: Testing the core-satellite species hypothesis. *Annales Zoologici Fennici* 19:65–73.

———. 1982c. Distributional ecology of anthropochorous plants in villages surrounded by forest. *Annales Botanici Fennici* 19:1–15.

———. 1997. Metapopulation dynamics: From concepts and observations to predictive models. In *Metapopulation biology,* ed. I. Hanski and M. E. Gilpin, 69–92. San Diego, CA: Academic Press.

Hanski, I., and M. E. Gilpin, eds. 1997. *Metapopulation biology.* San Diego, CA: Academic Press.

Hanski, I., J. Kouki, and A. Halkka. 1993. Three explanations of the positive relationship between distribution and abundance of species. In *Species diversity in ecological communities,* ed. R. E. Ricklefs and D. Schluter, 108–16. Chicago: University of Chicago Press.

Hanski, I., and D. Simberloff. 1997. The metapopulation approach, its history, conceptual domain, and application to conservation. In *Metapopulation biology,* ed. I. Hanski and M. E. Gilpin, 5–26. San Diego, CA: Academic Press.

Harvey, P. H., and J. H. Lawton. 1986. Patterns in three dimensions. *Nature* 324:212.

Harvey, P. H., and M. D. Pagel. 1991. *The comparative method in evolutionary biology.* Oxford: Oxford University Press.

Hassell, M. P., J. H. Lawton, and R. M. May. 1976. Patterns of dynamical behavior in single-species populations. *Journal of Animal Ecology* 45:471–86.

Hastings, A., C. L. Hom, S. Ellner, P. Turchin, and H. C. J. Godfray. 1993. Chaos in ecology: Is Mother Nature a strange attractor? *Annual Review of Ecology and Systematics* 24:1–33.

Hengeveld, R. 1990. *Dynamic biogeography.* Cambridge: Cambridge University Press.

Hengeveld, R., and J. Haeck. 1981. The distribution of abundance: II. Models and implications. *Proceedings of the Koninklijke Nederlandse Akademie van Wetenschappen,* series C 84:257–84.

———. 1982. The distribution of abundance: I. Measurements. *Journal of Biogeography* 9:303–16.

Hilborn, R., and M. Mangel. 1997. *The ecological detective: Confronting models with data.* Princeton, NJ: Princeton University Press.

Hoffman, A. 1989. *Arguments on evolution.* New York: Oxford University Press.

Holt, R. D. 1983. Immigration and the dynamics of peripheral populations. In *Advances in herpetology and evolutionary biology,* ed. A. G. J. Rhodin and K. Miyata, 680–94. Cambridge, MA: Museum of Comparative Zoology.

———. 1992. Theoretical ecology. In *McGraw-Hill encyclopedia of science and technology,* 7th ed., 3:272–74. New York: McGraw-Hill.

———. 1993. Ecology at the mesoscale: The influence of regional processes on local communities. In *Species diversity in ecological communities,* ed. R. E. Ricklefs and D. Schluter, 77–88. Chicago: University of Chicago Press.

———. 1996. Adaptive evolution in source-sink environments: Direct and indirect effects of density-dependence on niche evolution. *Oikos* 75:182–92.

———. 1997. From metapopulation dynamics to community structure: Some consequences of spatial heterogeneity. In *Metapopulation biology,* ed. I. Hanski and M. E. Gilpin, 149–64. San Diego, CA: Academic Press.

Holt, R. D., J. H. Lawton, K. J. Gaston, and T. M. Blackburn. 1997. On the relationship between range size and abundance: Back to the basics. *Oikos* 78: 183–90.

Hurlbert, S. H. 1984. Pseudoreplication and the design of ecological field experiments. *Ecological Monographs* 54:187–211.

Huston, M. A. 1994. *Biological diversity: The coexistence of species on changing landscapes.* Cambridge: Cambridge University Press.

Hutchinson, G. E. 1958. Concluding remarks. *Cold Spring Harbor Symposia on Quantitative Biology* 22:415–27.

———. 1959. Homage to Santa Rosalia or why are there so many kinds of animal? *American Naturalist* 93:145–59.

———. 1961. The paradox of the plankton. *American Naturalist* 95:137–45.

———. 1978. *An introduction to population ecology.* New Haven, CT: Yale University Press.

Hutchinson, G. E., and R. H. MacArthur. 1959. A theoretical ecological model of size distributions among species of animals. *American Naturalist* 93:117–25.

Ives, A. R. 1995. Predicting the response of populations to environmental change. *Ecology* 76:926–41.

Jackson, E. A. 1990. *Perspectives of nonlinear dynamics.* Cambridge: Cambridge University Press.

James, F. C., R. F. Johnston, N. O. Wamer, G. J. Niemi, and W. J. Boecklen. 1984. The Grinnellian niche of the wood thrush. *American Naturalist* 124:17–47.

Johnson, A. R., J. A. Wiens, B. T. Milne, and T. O. Crist. 1992. Animal movements and population dynamics in heterogeneous landscapes. *Landscape Ecology* 7:63–75.

Judson, O. P. 1994. The rise of the individual-based model in ecology. *Trends in Ecology and Evolution* 9:9–14.

Kareiva, P. 1994. Space: The final frontier for ecological theory. *Ecology* 75:1.

———. 1997. Why worry about the maturing of a science? EcoEssay Series no. 1. National Center for Ecological Analysis and Synthesis, Santa Barbara, CA.

Kauffman, S. A. 1993. *The origins of order: Self-organization and selection in evolution.* Oxford: Oxford University Press.

Kendall, D. G. 1948. On some modes of population growth leading to R. A. Fisher's logarithmic series distribution. *Biometrika* 35:6–15.

Kennett, J. P., and M. S. Srinvasan. 1983. *Neogene planktonic foraminifera: A phylogenetic atlas.* Stroudsberg, PA: Hutchinson Ross.

Kerner, E. H. 1957. A statistical mechanics of interacting biological species. *Bulletin of Mathematical Biophysics* 19:121–46.

———. 1959. Further consideration on the statistical mechanics of biological associations. *Bulletin of Mathematical Biophysics* 21:217–55.

Kingsland, S. E. 1995. *Modeling nature: Episodes in the history of population ecology.* Chicago: University of Chicago Press.

Kot, M., and W. M. Schaffer. 1984. The effects of seasonality on discrete models of population growth. *Theoretical Population Biology* 26:340–60.

Kotliar, N. B., and J. A. Wiens. 1990. Multiple scales of patchiness and patch structure: A hierarchical framework for the study of heterogeneity. *Oikos* 59: 253–60.

Kozłowski, J. 1996. Energetic definition of fitness? Yes, but not that one. *American Naturalist* 147: 1087–91.

Lack, D. 1947. *Darwin's finches.* Cambridge: Cambridge University Press.

Lack, D. L. 1971. *Ecological isolation in birds.* Cambridge, MA: Harvard University Press.

Lasota, A, and M. C. Mackey. 1994. *Chaos, fractals, and noise.* New York: Springer-Verlag.

Laszlo, E. 1987. *Evolution: The grand synthesis.* Boston: New Science Library.

Lawlor, L. R. 1979. Direct and indirect effects of *n*-species competition. *Oecologia* 43:355–64.

Lawton, J. H. 1989. What is the relationship between population density and body size in animals? *Oikos* 55:429–34.

Leibold, M. A. 1995. The niche concept revisited: Mechanistic models and community context. *Ecology* 76:1371–82.

Lele, S., M. L. Taper, and S. Gage. 1998. Statistical analysis of population dynamics in space and time using estimating functions. *Ecology,* in press.

Leopold, A. 1933. *Game management.* New York: Charles Scribner's Sons.

Levin, S. A. 1992. The problem of pattern and scale in ecology. *Ecology* 73:1943–67.

Levine, S. H. 1976. Competitive interactions in ecosystems. *American Naturalist* 110:903–10.

Levins, R. 1966. Strategy of model building in population biology. *American Scientist* 54:421–31.

———. 1968. *Evolution in changing environments.* Princeton, NJ: Princeton University Press.

———. 1973. The limits of complexity. In *Hierarchy theory: The challenge of complex systems,* ed. H. H. Pattee, 109–28. New York: George Braziller.

———. 1979. Coexistence in a variable environment. *American Naturalist* 114: 765–83.

Levins, R., and R. C. Lewontin. 1985. *The dialectical biologist.* Cambridge, MA: Harvard University Press.

Loehle, C. 1983. Evaluation of theories and calculation tools in ecology. *Ecological Modelling* 19:239–47.

Lomnicki, A. 1988. *Population ecology of individuals.* Princeton, NJ: Princeton University Press.

Lorenz, E. N. 1963. Deterministic nonperiodic flow. *Journal of Atmospheric Science* 20:130–41.

Lotka, A. J. 1920. Contribution to the general kinetics of material transformations. *Proceedings of the American Academy of Arts and Sciences* 55:137.

———. 1922a. Contribution to the energetics of evolution. *Proceedings of the National Academy of Sciences* 8:147–50.

———. 1922b. Natural selection as a physical principle. *Proceedings of the National Academy of Sciences* 8:151–55.

———. 1925. *Elements of mathematical biology.* New York: Williams & Wilkins.

Lubchenco, J., A. M. Olson, L. B. Brubaker, S. R. Carpenter, M. M. Holland, S. P. Hubbell, S. A. Levin, J. A. MacMahon, P. A. Matson, J. M. Melillo, H. A. Mooney, C. H. Peterson, H. R. Pulliam, L. A. Real, P. J. Regal, and P. G. Risser. 1991. The sustainable biosphere initiative: An ecological research agenda. *Ecology* 72:371–412.

Ludwig, J. A., and J. F. Reynolds. 1988. *Statistical ecology.* New York: Wiley.

MacArthur, R., and R. Levins. 1967. The limiting similarity, convergence, and divergence of coexisting species. *American Naturalist* 101:377–85.

MacArthur, R. H. 1958. Population ecology of some warblers of northeastern coniferous forests. *Ecology* 39:599–619.

———. 1969. Patterns of communities in the tropics. *Biological Journal of the Linnean Society* 1:19–30.

———. 1971. Patterns of terrestrial bird communities. In *Avian biology,* ed. D. S. Farner and J. R. King, 189–221. New York: Academic Press.

———. 1972. *Geographical ecology.* New York: Harper & Row.

MacArthur, R. H., and E. O. Wilson. 1967. *The theory of island biogeography.* Princeton, NJ: Princeton University Press.

Maddison, W. P., and D. R. Maddison. 1992. *MacClade: Analysis of phylogeny and character evolution.* Sunderland, MA: Sinauer Associates, Inc.

Magurran, A. E. 1988. *Ecological diversity and its measurement.* Princeton, NJ: Princeton University Press.

Maurer, B. A. 1985a. On the ecological and evolutionary roles of competition. *Oikos* 45:300–302.

———. 1985b. Avian community dynamics in desert grasslands: Observation scale and hierarchical structure. *Ecological Monographs* 55:295–312.

———. 1987. Scaling of biological community structure: A systems approach to community complexity. *Journal of Theoretical Biology* 127:97–110.

———. 1989. Diversity dependent species dynamics: Incorporating the effects of population level processes on species dynamics. *Paleobiology* 15:133–46.

Maurer, B. A. 1990a. The relationship between distribution and abundance in a patchy environment. *Oikos* 58:181–89.

———. 1990b. *Dipodomys* populations as energy processing systems: Regulation, competition, and hierarchical organization. *Ecological Modelling* 50:157–76.

———. 1994. *Geographical population analysis.* Oxford: Blackwell Scientific Publications.

———. 1996. Relating human population growth to the loss of biodiversity. *Biodiversity Letters* 3:1–5.

———. 1998a. The evolution of body size in birds, I: Evidence for nonrandom diversification. *Evolutionary Ecology,* in press.

———. 1998b. The evolution of body size in birds, II: The role of physiological constraint. *Evolutionary Ecology,* in press.

Maurer, B. A., and J. H. Brown. 1988. Distribution of energy use and body mass among species of North American terrestrial birds. *Ecology* 69:1923–32.

———. 1989. Distributional consequences of spatial variation in local demographic processes. *Annales Zoologici Fennici* 26:121–31.

Maurer, B. A., J. H. Brown, and R. D. Rusler. 1992. The micro and macro in body size evolution. *Evolution* 46:939–53.

Maurer, B. A., H. A. Ford, and E. H. Rapoport. 1991. Extinction rate, body size, and avifaunal diversity. *Acta XX Congressus Internationalis Ornithologici* (Wellington, New Zealand) 2:826–34.

Maurer, B. A., and M. P. Nott. 1998. Geographic range fragmentation and the evolution of biological diversity. In *Biodiversity dynamics: Turnover of popula-*

tions, species, higher taxa, and communities, ed. M. L. McKinney. New York: Columbia University Press, forthcoming.

Maurer, B. A., and M.-A. Villard. 1994. Geographic variation in abundance of North American birds. *Research and Exploration* 10:307–17.

May, R. M. 1974. *Stability and complexity in model ecosystems.* Princeton, NJ: Princeton University Press.

———. 1975. Biological populations obeying difference equations: Stable points, stable cycles, and chaos. *Journal of Theoretical Biology* 51:511–24.

———. 1976. Simple mathematical models with very complicated dynamics. *Nature* 261:459–67.

———. 1978. The dynamics and diversity of insect faunas. In *Diversity of insect faunas,* ed. L. A. Mound and N. Waloff, 188–204. London: Blackwell Scientific Publications.

———. 1986. The search for patterns in the balance of nature: Advances and retreats. *Ecology* 67:1115–26.

May, R. M., and G. F. Oster. 1976. Bifurcations and dynamic complexity in simple ecological models. *American Naturalist* 110:573–99.

Maynard Smith, J. 1974. *Models in ecology.* Cambridge: Cambridge University Press. 110:331–38.

———. 1989. *Evolutionary genetics.* Oxford: Oxford University Press.

Mayr, E. 1961. Cause and effect in biology. *Science* 134:1501–6.

McCulloch, C. E. 1985. Variance tests for species association. *Ecology* 66:1676–81.

McGuinness, K. A. 1984. Equations and explanations in the study of species-area relationships. *Biological Reviews* 59:423–40.

McIntosh, R. P. 1985. *The background of ecology: Concept and theory.* Cambridge: Cambridge University Press.

McShea, D. W. 1994. Mechanisms of large-scale evolutionary trends. *Evolution* 48:1747–63.

Menge, B. A. 1997. Detection of direct versus indirect effects: Were experiments long enough? *American Naturalist* 149:801–23.

Menge, B. A., and A. M. Olson. 1990. Role of scale and environmental factors in regulation of community structure. *Trends in Ecology and Evolution* 5:52–57.

Milliken, G. A., and D. E. Johnson. 1992. *Analysis of messy data, volume I: Designed experiments.* New York: Chapman & Hall.

Morse, D. R., J. H. Lawton, M. M. Dodson, and M. H. Williamson. 1985. Fractal dimension of vegetation and the distribution of arthropod body lengths. *Nature* 314:731–33.

Morse, D. R., N. E. Stork, and J. H. Lawton. 1988. Species number, species abundance and body length relationships of arboreal beetles in Bornean lowland rain forest trees. *Ecological Entomology* 13:25–37.

Nagy, K. A. 1987. Field metabolic rate and food requirement scaling in mammals and birds. *Ecological Monographs* 57:111–28.

Nowak, R. M. 1991. *Walker's mammals of the world.* Baltimore, MD: Johns Hopkins University Press.

O'Neill, R. V. 1989. Perspectives in hierarchy and scale. In *Perspectives in ecologi-*

cal theory, ed. J. Roughgarden, R. M. May, and S. A. Levin, 140–56. Princeton, NJ: Princeton University Press.

O'Neill, R. V., D. L. DeAngelis, J. B. Waide, and T. F. H. Allen. 1986. *A hierarchical concept of ecosystems.* Princeton, NJ: Princeton University Press.

Odum, E. P. 1969. The strategy of ecosystem development. *Science* 164:262–70.

Odum, H. T. 1994. *Ecological and general systems and introduction to systems ecology.* Niwot: University Press of Colorado.

Osenberg, C. W., O. Sarnelle, and S. D. Cooper. 1997. Effect size in ecological experiments: The application of biological models in meta-analysis. *American Naturalist* 150:798–812.

Owen, J., and F. S. Gilbert. 1989. On the abundance of hoverflies. *Oikos* 55: 183–93.

Pace, M. L. 1993. Forecasting ecological responses to global change: The need for large-scale comparative studies. In *Biotic interactions and global change,* ed. P. M. Kareiva, J. G. Kingsolver, and R. B. Huey, 356–63. Sunderland, MA: Sinauer Associates, Inc.

Pack, N. 1996. Scale-sensitive linear analysis of granivorous rodent population dynamics. Master's thesis, Brigham Young University.

Pahl-Wostl, C. 1995. *The dynamic nature of ecosystems: Chaos and order intertwined.* New York: Wiley.

Parker, T. S., and L. O. Chua. 1989. *Practical numerical algorithms for chaotic systems.* New York: Springer-Verlag.

Patterson, B. D., and W. Atmar. 1986. Nested subsets and the structure of insular mammalian faunas and archipelagos. *Biological Journal of the Linnean Society* 28:65–82.

Peitgen, H.-O., J. Jürgens, and D. Saupe. 1992. *Chaos and fractals: New frontiers in science.* New York: Springer-Verlag.

Perrin, N., and R. M. Sibly. 1993. Dynamic models of energy allocation and investment. *Annual Review of Ecology and Systematics* 24:379–410.

Peters, R. H. 1976. Tautology in ecology and evolution. *American Naturalist* 110: 1–12.

———. 1983. *The ecological implications of body size.* Cambridge: Cambridge University Press.

———. 1991. *A critique for ecology.* Cambridge: Cambridge University Press.

Pickett, S. T. A., J. Kolasa, and C. G. Jones. 1994. *Ecological understanding: The nature of theory and the theory of nature.* San Diego, CA: Academic Press.

Pickett, S. T. A., and P. White, eds. 1985. *The ecology of natural disturbance and patch dynamics.* New York: Academic Press.

Pielou, E. C. 1975. *Ecological diversity.* New York: Wiley.

———. 1977. *Mathematical ecology.* New York: Wiley.

———. 1981. The usefulness of ecological models: A stock-taking. *Quarterly Review of Biology* 56:17–31.

———. 1984. *The interpretation of ecological data: A primer on classification and ordination.* New York: Wiley.

Pimm, S. L. 1979. Sympatric speciation: A simulation study. *Biological Journal of the Linnean Society* 11:131–39.

———. 1982. *Food webs.* London: Chapman & Hall.
———. 1991. *The balance of nature?* Chicago: University of Chicago Press.
Pimm, S. L., and A. Redfearn. 1988. The variability of animal populations. *Nature* 334:613–14.
Powell, T. M. 1989. Physical and biological scales of variability in lakes, estuaries, and the coastal ocean. In *Perspectives in ecological theory,* ed. J. Roughgarden, R. M. May, and S. A. Levin, 157–76. Princeton, NJ: Princeton University Press.
Preston, F. W. 1948. The commonness, and rarity, of species. *Ecology* 29:254–83.
———. 1962a. The canonical distribution of commonness and rarity: Part I. *Ecology* 43:185–215.
———. 1962b. The canonical distribution of commonness and rarity: Part II. *Ecology* 43:410–32.
Price, J., S. Droege, and A. Price. 1995. *The summer atlas of North American birds.* San Diego, CA: Academic Press.
Prigogine, I., and I. Stengers. 1984. *Order out of chaos: Man's new dialogue with nature.* New York: Bantam.
Puccia, C. J., and R. Levins. 1985. *Qualitative modeling of complex systems.* Cambridge, MA: Harvard University Press.
Pulliam, H. R. 1975. Coexistence of sparrows: A test of community theory. *Science* 189:474–76.
———. 1985. Foraging efficiency, resource partitioning, and the coexistence of sparrow species. *Ecology* 66:1829–36.
———. 1988. Sources, sinks, and population regulation. *American Naturalist* 132:652–61.
Quinn, J. F., and A. Hastings. 1987. Extinction in subdivided habitats. *Conservation Biology* 1:198–208.
———. 1988. Extinction in subdivided habitats: Reply to Gilpin. *Conservation Biology* 2:293–96.
Rapoport, E. 1982. *Areography: Geographical strategies of species.* Oxford: Pergamon.
Rencher, A. C. 1995. *Methods of multivariate analysis.* New York: Wiley.
Ricklefs, R. E. 1974. Energetics of reproduction in birds. In *Avian energetics,* ed. R. A. Paynter, 152–297. Cambridge, MA: Nuttall Ornithological Club.
———. 1987. Community diversity: Relative roles of local and regional processes. *Science* 235:167–71.
Ricklefs, R. E., and R. E. Latham. 1992. Intercontinental correlation of geographical ranges suggests stasis in ecological traits of relict genera of temperate perennial herbs. *American Naturalist* 139:1305–21.
Ricklefs, R. E., and D. Schluter, eds. 1993. *Species diversity in ecological communities.* Chicago: University of Chicago Press.
Robbins, C. S., D. Bystrak, and P. H. Geissler. 1986. *The Breeding Bird Survey: Its first fifteen years.* Resource Publication no. 157. United States Department of the Interior Fish and Wildlife Service, Washington, D.C.
Robinson, J. G., and K. H. Redford. 1986. Body size, diet, and population density of neotropical forest mammals. *American Naturalist* 128:665–80.
Roff, D. A. 1992. *The evolution of life histories.* New York: Chapman & Hall.

Root, T. 1988a. Environmental factors associated with avian distributional boundaries. *Journal of Biogeography* 15:489–505.
———. 1988b. Energy constraints on avian distributions and abundances. *Ecology* 69:330–39.
———. 1988c. *Atlas of wintering North American birds.* Chicago: University of Chicago Press.
Root, T. L., and S. H. Schneider. 1995. Ecology and climate: Research strategies and implications. *Science* 269:334–41.
Rosenzweig, M. L. 1975. On continental steady states of species diversity. In *Ecology and evolution of communities,* ed. M. L. Cody and J. M. Diamond, 121–40. Cambridge, MA: Belknap Press.
———. 1978. Competitive speciation. *Biological Journal of the Linnean Society* 10:275–89.
———. 1995. *Species diversity in space and time.* Cambridge: Cambridge University Press.
Roughgarden, J. 1979. *Theory of population genetics and evolutionary ecology: An introduction.* New York: Macmillan.
Roughgarden, J., S. D. Gaines, and S. W. Pacala. 1987. Supply side ecology: The role of physical transport processes. In *Organization of communities past and present: The 27th symposium of the British ecological society,* ed. J. H. R. Gee and P. S. Giller, 491–518. Oxford: Blackwell Scientific Publications.
Royama, T. 1992. *Analytical population dynamics.* London: Chapman & Hall.
Ruelle, D. 1989. *Chaotic evolution and strange attractors.* Cambridge: Cambridge University Press.
———. 1991. *Chance and chaos.* Princeton, NJ: Princeton University Press.
Sale, P. F. 1984. The structure of communities of fish on coral reefs and the merit of a hypothesis-testing, manipulative approach to ecology. In *Ecological communities: Conceptual issues and the evidence,* ed. D. R. Strong, D. Simberloff, L. G. Abele, and A. B. Thistle, 478–90. Princeton, NJ: Princeton University Press.
Salthe, S. N. 1985. *Evolving hierarchical systems.* New York: Columbia University Press.
———. 1989. Self-organization of/in hierarchically structured systems. *Systems Research* 6:199–208.
Schaffer, W. M. 1981. Ecological abstraction: The consequences of reduced dimensionality in ecological models. *Ecological Monographs* 51:383–401.
———. 1985. Order and chaos in ecological systems. *Ecology* 66:93–106.
Schluter, D. 1984. A variance test for detecting species associations, with some example applications. *Ecology* 65:998–1005.
———. 1986. Tests for similarity and convergence of finch communities. *Ecology* 67:1073–85.
Schneider, S. H., and P. J. Boston, eds. 1991. *Scientists on Gaia.* Cambridge, MA: MIT Press.
Schoener, T. W. 1974. Resource partitioning in ecological communities. *Science* 185:27–39.
———. 1982. The controversy over interspecific competition. *American Scientist* 70:586–95.

———. 1983. Field experiments on interspecific competition. *American Naturalist* 122:240–85.
———. 1986. Overview: Kinds of ecological communities—ecology becomes pluralistic. In *Community ecology,* ed. J. Diamond and T. J. Case, 467–79. New York: Harper & Row.
———. 1989. Food webs from the small to the large. *Ecology* 70:1559–89.
Schrödinger, E. 1944. *What is life?* Cambridge: Cambridge University Press.
Scott, J. M., F. Davis, B. Csuti, R. Noss, B. Butterfield, C. Groves, H. Anderson, S. Caicco, F. D'Erchia, T. C. Edwards, J. Ulliman, and R. G. Wright. 1993. Gap analysis: A geographic approach to protection of biological diversity. *Wildlife Monographs* 123:1–41.
Sepkoski, J. J., Jr. 1978. A kinetic model of Phanerozoic taxonomic diversity: I. Analysis of marine orders. *Paleobiology* 4:223–51.
Shaw, R. 1981. Strange attractors, chaotic behavior, and information flow. *Zeitschrift fuer Naturforschung* 36A:80–112.
Short, L. L., and J. B. Hestbeck. 1995. National biotic resource inventories and GAP analysis. *BioScience* 45:535–39.
Sibley, C. G., J. E. Alquist, and B. L. Monroe, Jr. 1988. A classification of the living birds of the world based on DNA-DNA hybridization studies. *Auk* 105: 409–23.
Sibley, C. G., and B. L. Monroe, Jr. 1990. *Distribution and taxonomy of birds of the world.* New Haven, CT: Yale University Press.
Sih, A., P. Crowley, M. McPeek, J. Petranka, and K. Strohmeier. 1985. Predation, competition, and prey communities: A review of field experiments. *Annual Review of Ecology and Systematics* 16:269–311.
Simberloff, D. 1982. The status of competition theory in ecology. *Annales Zoologici Fennici* 19:241–53.
———. 1983. Competition theory, hypothesis-testing, and other community ecological buzzwords. *American Naturalist* 122:626–35.
Smith, F. A., J. H. Brown, and T. J. Valone. 1997. Path analysis: A critical evaluation using long-term experimental data. *American Naturalist* 149:29–42.
Smolin, L. 1997. *The life of the cosmos.* New York: Oxford University Press.
Sommerer, J. C., and E. Ott. 1993. A physical system with qualitatively uncertain dynamics. *Nature* 365:138–40.
Stanley, S. M. 1973. An explanation for Cope's rule. *Evolution* 27:1–26.
———. 1979. *Macroevolution.* San Francisco, CA: Freeman, Cooper & Co.
———. 1982. Macroevolution and the fossil record. *Evolution* 36:460–73.
———. 1986. Population size, extinction, and speciation: The fission effect in Neogene Bivalvia. *Paleobiology* 12:89–110.
———. 1990. The general correlation between rate of speciation and rate of extinction: Fortuitous causal linkages. In *Causes of evolution: A paleontological perspective,* ed. R. M. Ross and W. D. Allmon, 103–27. Chicago: University of Chicago Press.
Stanley, S. M., K. L. Wetmore, and J. P. Kennett. 1988. Macroevolutionary differences between the two major clades of Neogene planktonic foraminifera. *Paleobiology* 14:235–49.
Stauffer, R. C. 1975. *Charles Darwin's natural selection.* Cambridge: Cambridge University Press.

Stearns, S. C. 1992. *The evolution of life histories.* Oxford: Oxford University Press.
Stenseth, N. C. 1985. Darwinian evolution in ecosystems: The Red Queen view. In *Evolution,* ed. P. J. Greenwood, P. H. Harvey, and M. Slatkin, 55–72. Cambridge: Cambridge University Press.
Stenseth, N. C., and J. Maynard Smith. 1984. Coevolution in ecosystems: Red Queen evolution or stasis? *Evolution* 38:870–80.
Stone, L., T. Dayan, and D. Simberloff. 1996. Community-wide assembly patterns unmasked: The importance species' differing geographical ranges. *American Naturalist* 148:997–1015.
Stone, L., and A. Roberts. 1991. Conditions for a species to gain advantage from the presence of a competitor. *Ecology* 72:1964–72.
Strayer, D. 1986. The size structure of a lacustrine zoobenthic community. *Oecologia* 69:513–16.
Strong, D. R., Jr. 1983. Natural variability and the manifold mechanisms of ecological communities. *American Naturalist* 122:636–60.
Sugihara, G. 1980. Minimal community structure: An explanation of species abundance patterns. *American Naturalist* 116:770–87.
Sugihara, G., G. Grenfell, and R. M. May. 1990. Distinguishing error from chaos in ecological time series. *Philosophical Transactions of the Royal Society of London,* series B 330:235–51.
Sugihara, G., and R. M. May. 1990a. Applications of fractals in ecology. *Trends in Ecology and Evolution* 5:79–86.
———. 1990b. Nonlinear forecasting as a way of distinguishing chaos from measurement error in time series. *Nature* 344:734–41.
Thompson, J. N. 1994. *The coevolutionary process.* Chicago: University of Chicago Press.
Tilman, D. 1982. *Resource competition and community structure.* Princeton, NJ: Princeton University Press.
———. 1988. Plant strategies and the dynamics and structure of plant communities. Princeton, NJ: Princeton University Press.
Turchin, P., and A. D. Taylor. 1992. Complex dynamics in ecological time series. *Ecology* 73:289–305.
Turner, B. L., W. C. Clark, R. W. Kates, J. F. Richards, J. T. Matthews, and W. B. Meyer. 1990. *The earth as transformed by human action.* Cambridge: Cambridge University Press.
Turner, M. G., R. H. Gardner, and R. V. O'Neill. 1995. Ecological dynamics at broad scales: Ecosystems and landscapes. *Bioscience Supplement: Science & Biodiversity Policy* S29–S35.
Ulanowicz, R. E. 1986. *Growth and development: Ecosystems phenomenology.* New York: Springer-Verlag.
———. 1997. *Ecology, the ascendent perspective.* New York: Columbia University Press.
Underwood, A. J. 1986. The analysis of competition by field experiments. In *Community ecology: Pattern and process,* ed. J. Kikkawa and D. J. Anderson, 240–68. Boston: Blackwell Scientific Publications.
———. 1997. *Experiments in ecology.* Cambridge: Cambridge University Press.
Urban, D. L., R. V. O'Neill, and H. H. Shugart. 1987. Landscape ecology. *BioScience* 37:119–27.

Valone, T. J., and J. H. Brown. 1995. Effects of competition, colonization, and extinction on rodent species. *Science* 267:880–83.

Van Horne, B. 1983. Density as a misleading indicator of habitat quality. *Journal of Wildlife Management* 47:893–901.

Vitousek, P. M. 1994. Beyond global warming: Ecology and global change. *Ecology* 75:1861–77.

Vitousek, P. M., P. R. Ehrlich, A. H. Ehrlich, and P. A. Mateson. 1986. Human appropriation of the products of photosynthesis. *BioScience* 36:368–73.

Vitousek, P. M., H. A. Mooney, J. Lubchenco, and J. M. Melillo. 1997. Human domination of the earth's ecosystems. *Science* 277:494–99.

Volterra, V. 1926. Fluctuations in the abundance of a species considered mathematically. *Nature* 118:558–60.

Vrba, E. S. 1980. Evolution, species and fossils: How does life evolve? *South African Journal of Science* 76:61–84.

Walker, T. D. 1985. Diversification functions and the rate of taxonomic evolution. In *Phanerozoic diversity patterns: Profiles in macroevolution,* ed. J. W. Valentine, 311–34. Princeton, NJ: Princeton University Press.

Wei, W. S. 1990. *Time series analysis: Univariate and multivariate methods.* New York: Addison-Wesley.

Whittaker, R. H. 1967. Gradient analysis of vegetation. *Biological Reviews* 42: 207–69.

———. 1975. *Communities and ecosystems.* 2d ed. New York: Macmillan.

Wiens, J. A. 1977. On competition and variable environments. *American Scientist* 65:590–97.

———. 1983. Avian community ecology: An iconoclastic view. In *Perspectives in ornithology,* ed. A. H. Brush and G. A. Clark, Jr., 355–403. Cambridge: Cambridge University Press.

———. 1984. On understanding a non-equilibrium world: Myth and reality in community patterns and processes. In *Ecological communities: Conceptual issues and the evidence,* ed. D. R. Strong, D. Simberloff, L. G. Abele, and A. B. Thistle, 439–57. Princeton, NJ: Princeton University Press.

———. 1986. Spatial scale and temporal variation in studies of shrubsteppe birds. In *Community ecology,* ed. J. Diamond and T. J. Case, 154–72. New York: Harper & Row.

———. 1989a. Spatial scaling in ecology. *Functional Ecology* 3:385–97.

———. 1989b. *The ecology of bird communities.* 2 vols. Cambridge: Cambridge University Press.

———. 1995. Landscape mosaics and ecological theory. In *Mosaic landscapes and ecological processes,* ed. L. Hansson, L. Fahrig, and G. Merriam, 1–26. London: Chapman & Hall.

Wiens, J. A., J. F. Addicott, T. J. Case, and J. Diamond. 1986. Overview: The importance of spatial and temporal scale in ecological investigations. In *Community ecology,* ed. J. Diamond and T. J. Case, 145–53. New York: Harper & Row.

Wiens, J. A., N. C. Stenseth, B. Van Horne, and R. A. Ims. 1993. Ecological mechanisms and landscape ecology. *Oikos* 66:369–80.

Williams, C. B. 1964. *Patterns in the balance of nature.* New York: Academic Press.

Williamson, M. 1981. *Island populations.* Oxford: Oxford University Press.

Wilson, D. S. 1983. The group selection controversy: History and current status. *Annual Review of Ecology and Systematics* 14:159–87.

———. 1986. Adaptive indirect effects. In *Community ecology,* ed. J. Diamond and T. J. Case, 437–44. New York: Harper & Row.

Wilson, E. O. 1992. *The diversity of life.* New York: Norton.

Wing, L. 1943. Spread of the starling and English sparrow. *Auk* 60:74–87.

Wootton, J. T. 1994. Putting the pieces together: Testing independence of interactions among organisms. *Ecology* 75:1544–51.

Wright, D. H. 1990. Human impacts on energy flow through natural ecosystems, and implications for species endangerment. *Ambio* 19:189–94.

———. 1991. Correlations between incidence and abundance are expected by chance. *Journal of Biogeography* 18:463–66.

Wright, D. H., and J. H. Reeves. 1992. On the meaning and measurement of nestedness of species assemblages. *Oecologia* 92:416–28.

Wright, S. J. 1931. Evolution in Mendelian populations. *Genetics* 16:97–159.

Yeaton, R. I. 1974. An ecological analysis of chaparral and pine forest bird communities on Santa Cruz Island and mainland California. *Ecology* 55: 959–73.

Yodzis, P. 1988. The indeterminacy of ecological interactions as perceived through perturbation experiments. *Ecology* 69:508–15.

———. 1989. *Introduction to theoretical ecology.* New York: Harper & Row.

Yodzis, P., and S. Innes. 1992. Body size and consumer-resource dynamics. *American Naturalist* 139:1151–75.

INDEX